ii

Contents

1—22811

APPENDICES

DISTRIBUTION

Infantry .. .. Scale A plus 10
Small Arms School (Platoon Weapons) .. .. 200 copies
School of Infantry .. .. 50 copies

ABBREVIATIONS

A.P.	Armour piercing.
B.L.R.	Beyond local repair.
D.F.	Defensive fire.
E.M.E.R.	Electrical and Mechanical Engineer Regulation.
F.D.L.	Forward defended locality.
I.O.	Intelligence officer.
M.P.I.	Mean point of impact.
O.P.	Observation post.
S.A.A.	Small arms ammunition.

Small Arms Training
Volume I, Pamphlet No. 28

Sniping   Part I

Introduction

1. Opportunities will present themselves in all types of warfare for the expert use of the rifle, as an offensive weapon. Although the rifle is still the personal weapon of the majority of infantrymen, special training is required for its offensive use as an independent fire unit. This training incorporates a wide variety of subjects, which are as important as skill with the weapon itself; and the value of such specialists, or snipers, has been found to extend in many directions, beyond their primary task of shooting.

The subject of sniping, therefore, though essentially dealing with the use of the rifle, by the individual expert, must also embrace some of these important bye-products.

Chapter 1

Section 1   Organization

2. The War Establishment of an infantry battalion provides for a sniping section, on battalion headquarters, consisting of :—

> 1 Serjeant.
> 1 Corporal.
> 2 Lance-Corporals.
> 4 Privates.

3. Although no specialist sniping officer exists on the battalion war establishment, it is most essential that some officer having a thorough knowledge of the subject should supervise their selection, training and handling in battle. On paper the intelligence officer is responsible, among his other duties, for them, but in practice some other officer is often entrusted with these tasks.

On the enthusiasm of the officer in question, on his judgment and skill in choosing his team, and on his energy in directing their training and operational tasks, the measure of their success depends. Whatever their individual prowess, it has been found that this will be wasted, without such leadership and support.

4. As for all other specialists, it is necessary to train 100 per cent. reserves of snipers. In battle, it may often be found necessary, in order to provide reliefs, to employ these reserves operationally in conjunction with the first line snipers.

Section 2   Equipment

5. The authorized scale of specialist equipment for snipers, with an infantry battalion is :—

Rifles No. 4, Mark 1 (T) (*i.e.*, fitted with a No. 32 telescopic sight)	8
Telescopes, Scout Regiment, Mark 2	8
Binoculars No. 2, prismatic	8
Smocks, Denison	16
Watches, G.S.	8
Compasses, Prismatic, Liquid, Mark 3	8

In addition, one extra face veil is authorized for each sniper ; and this is expendable.

Snipers will normally operate in pairs, and this scale, it will be seen, is sufficient to equip 8 pairs, if it should be decided to employ the reserves operationally.

Equipment of a sniper

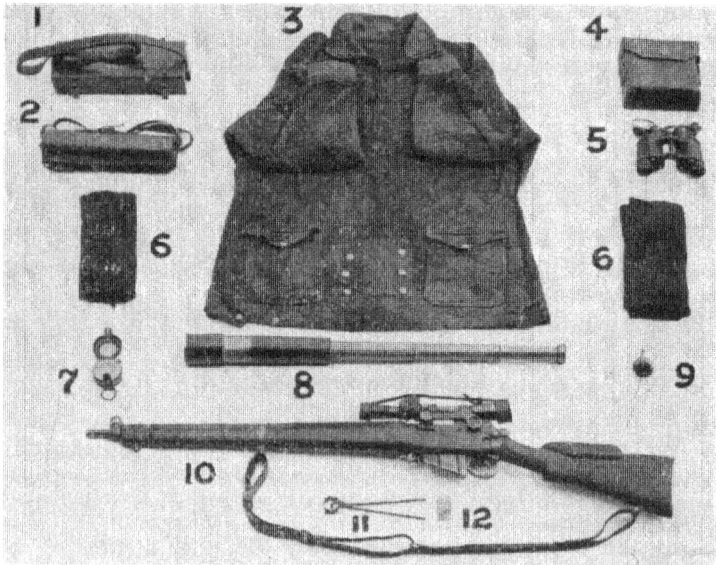

1. Case for telescope, sighting, No. 32	7. Compass, liquid, prismatic
2. Case for telescope, scout regiment	8. Telescope, scout regiment
3. Denison smock	9. Watch. G.S.
4. Case for binoculars	10. Rifle No. 4, Mark 1 (T) with telescope sighting No. 32
5. Binoculars, prismatic	11. Tool adjusting No. 1, Mark 1
6. Face veils	12. Tool adjusting No. 2, Mark 1

6. With the pair, a machine carbine, or an ordinary rifle, must be carried by the man not having the sniper rifle, and the following extra items of equipment have been found practicable :—

    50 rounds S.A.A.
    2 grenades (No. 36 or No. 77).
    5 rounds tracer (kept separate from the ball ammunition).
    5 rounds A.P.
    1 water bottle.
    1 emergency ration.

These items are liable to much variation, in accordance with the task in hand. But the maxim must always be to travel as lightly as possible. To this end the Denison smock, with its ample pockets, is a most useful garment, because it eliminates the need for web equipment.

Section 3   Characteristics and training of snipers

7. A fully trained sniper can be defined as a soldier, who is trained to locate an enemy, however well hidden, who can stalk or lie in wait for him unseen, and who is an expert shot with the rifle. His object is to kill with one round. He is the big game hunter of the battlefield, and must combine the art of the hunter, the wiles of a poacher and the skill of a target shot, with the determination to seek out his enemy.

8. He must be trained not only in rifle markmanship, but also to a high degree in observation and in fieldcraft. Without great skill as an observer, he will seldom be able to find suitable targets ; and without a comprehensive knowledge of fieldcraft, he will rarely get to a fire position within shot of his quarry ; thus observation and fieldcraft must be thoroughly understood by him.

9. In addition, there is a large number of subjects, which will be of great value towards making him a reliable and self-reliant hunter. Many of them form a part of the three main subjects given above, but some of the additional ones are map reading, compass work, photo reading, reporting, and a knowledge of enemy weapons, identifications and minor tactics.

There can be no real end to a sniper's training, in the hands of an imaginative instructor, which partly accounts for snipers being centralized in battalion headquarters, where their specialist training can be continuous.

10. The sniper's task is to kill individual enemy with single rifle shots very quickly aimed if necessary. He will never fire a rapid suc-cession of shots, except in self-defence. As a guide, the standard of shooting to be demanded of a sniper will enable him to hit, with security, a man's head up to 200 yards, and a man's trunk up to 300 yards, though this standard may well be surpassed. Under favourable weather conditions, a chance may exist of hitting a man's body up to 1,000 yards ; ~~but shooting at anything approaching such a range must be strongly discouraged, unless there is some very special reason for doing so~~.

11. Where a choice of targets exists, the sniper should be trained to pick out, carefully, the most important ones, such as enemy officers and N.C.Os., who can often be recognized by their actions, as well as the badges they may carry. Observers, snipers, signallers and

2a—22811 Delete from "but" in line 8 to end of paragraph and substitute "and when it is impossible to stalk closer to the enemy, due to the nature of the ground, and the only alternative is to do nothing, the sniper should be able to take on such targets provided that he can ascertain the exact range accurately from his map or air photograph"

wireless men, runners, M.G. and mortar teams, or any man engaged on some special task, are all targets to be sought after, and their loss to the enemy will be out of all proportion to the number of rounds fired. Other targets that can be effectively dealt with by accurate rifle fire are loopholes or embrasures, which are often proof against the fire of high trajectory weapons. Enemy weapons have themselves been put out of action, on occasions, by an A.P. bullet, fired into their mechanisms ; such action, where possible, is more effective than to deal, one by one, with the team that mans the weapon.

12. The light-collecting properties of the telescopic sight which are one of its main advantages, permit of accurate fire under conditions of half light, when iron sights would be useless. The best results at sniping are often obtained on moonlight nights and at dusk and dawn, at which times many tasks are carried out, and the enemy may be lulled into a sense of security against aimed fire, which facts should therefore be exploited fully.

13. The special observation training of a sniper, as well as the nature of his work, result in him making an important contribution to intelligence. If he makes efficient use of the observer telescope, which has a very high power, he will normally see more than any other form of O.P. He must, therefore, be trained, always to make a report after carrying out any task. Snipers have often been of great assistance to gunner O.Ps., and occasionally it may be worth giving them signal communications. It must be emphasized, however, that the primary task of snipers is to seek out and kill the enemy, and excepting where there is no possibility at all of doing this, it is wrong to use them purely for manning an O.P., which is normally the task of the intelligence section.

14. The training of the sniper in fieldcraft will make him a valuable soldier for many tasks other than sniping alone, but the tendency to use him for all sorts of patrolling and scouting tasks is to be guarded against, and should not be resorted to, where such jobs could be done by less skilled men. Otherwise, since these tasks are apt to be very strenuous, he will be frequently not available for his proper role. On the other hand, bold and original uses of snipers, on opportunity tasks, can, if properly planned, yield high dividends, particularly against a retreating and disorganized enemy, and where the country lends itself to infiltration.

15. Snipers normally achieve the best results when operating in pairs, for the following reasons :—

(a) One man cannot observe and shoot at the same time, and it is a great help for him to have an observer to find and indicate targets to him.

(b) Observation, especially with the telescope, involves great strain and can only be continued for limited periods. The pair must, therefore, be interchangeable.

(c) It is seldom possible for the firer to observe the strike of his shot. The observer will generally be able to do so, and give any correction required.

(d) Mutual protection is afforded, especially during movement.

(e) The moral effect of companionship in an otherwise lonely task is considerable.

Against the above it will be apparent that when stalking the enemy at close range, one man will have less chance of being seen than two ; and on such occasions, it is often better for the sniper to proceed alone, possibly with his partner in a covering fire position to his rear.

Section 4   Selection of snipers

16. In order to get good results from sniping great care is necessary in selecting the men for the task. Experience has shown that the primary consideration is a self-reliant personality, with a determination to seek out the enemy, and a liking for the work. Without these qualities the value of any others will be small ; and it is especially emphasized that skill as a shot is not in itself sufficient. The following points, however, require attention.

(a) Eyesight and night vision must be good.

(b) A natural aptitude for fieldcraft is necessary. This and good eyesight are generally to be found in countrymen, and especially in keepers, stalkers and poachers.

(c) Intelligence and initiative are required, since snipers work independently, to a large extent, and must be trained in a wide variety of subjects.

(d) Promise as a shot is essential, since, unless he reaches a high degree of skill with the rifle, the rest of the sniper's training is purposeless. Provided a man has normal physical and mental attributes and can group to within 6 inches at 100 yards, with iron sights, careful coaching will generally bring him up to the required standard.

Section 5   Tactical handling of snipers

17. The best results from sniping are only achieved when the employment of snipers is carefully organized. Despite the independent nature of their work, it has been found that little is achieved, and casualties among the snipers themselves are unnecessarily high, unless this is done. Thus sniping tasks will normally be planned on a brigade level ; though they may sometimes be decentralized to companies for specific purposes.

18. Snipers are largely a weapon of opportunity, and broad principles only can be given as to their use. These principles must be applied with imagination, guided by a sound grasp of snipers' characteristics.

Section 6   Mobile warfare

19. When conditions are mobile, snipers can often stalk and shoot isolated enemy posts that cannot be located or are difficult to deal with by other means. Isolated pillboxes, or M.G. posts holding up an advance, have, on many occasions, been overcome in this way.

20. Under favourable circumstances, infiltration of snipers through the enemy positions may be very successful ; and when this is done before an attack, they may give great assistance, by shooting enemy

weapon crews from the rear. On some occasions they have even been able to interfere, in this way, with the enemy mortar positions, or gun lines. When snipers are sent out on such tasks it is necessary to take precautions against their becoming involved in our own fire plans.

21. When interference is met with from enemy snipers left to delay our advance, our own snipers may be of assistance in the role of counter-sniping. The use of snipers by a retreating enemy is frequently resorted to in close or wooded country, and the greatest success against them is usually gained by hunting them with small parties consisting of snipers or with snipers attached.

Section 7   Static conditions

22. When a battle becomes stabilized, the prime object of sniping is to dominate the no-man's-land between the opposing forces. This entails the prevention of enemy sniping, or front line observation, and interference with his freedom of movement and patrolling activity.

23. The result of such a policy, if carried out offensively by well-trained and well-organized snipers, will do more than inflict casualties and inconvenience on the enemy. It will also have a marked effect on the security and morale of our own troops, and will result in a constant source of information about the enemy. Conversely, it will have a demoralizing effect on the enemy.

24. The methods by which this object may be achieved will be governed by many factors, such as the nature of the ground, the distance between the opposing F.D.Ls., the existence of minefields or other obstacles, the degree of initiative held by the enemy, and the number of trained snipers available.

25. When the enemy positions are within the range for accurate rifle fire, snipers will normally operate within our own F.D.Ls, from hides, or prepared positions, which must be carefully sited, so as to cover, as far as possible, by interlapping arcs, the whole of the battalion front. They must also be most carefully concealed and they should not be sited close to company positions or O.Ps.

26. When the distance between the opposing F.D.Ls. is too great for rifle shooting, sniping posts, with good fields of view, should be established forward of our F.D.Ls. This will enable snipers to hamper or put a stop to enemy patrolling and other activities in no-man's-land. It will also provide a system of forward O.Ps., which will gain useful observations, and add to the security of the main positions. Often it will only be possible to get snipers to and from these posts under cover of darkness. Where this necessitates the snipers remaining in the post through the hours of daylight, great strain is put on the men themselves, and the question of reliefs needs careful consideration.

27. The posts should be reconnoitred and allocated personally by the officer in charge of the snipers, whose responsibility it is also to ensure that they do not come within any of our own D.F. tasks, which may have to be altered to fit in with sniping operations. Full information about our own and enemy troops and minefields should be given to the snipers, and they should be briefed in the manner of a patrol. All troops must be warned of the operation of snipers on their front, or

casualties among the snipers will occur from our own fire. On the other hand it is often possible for fire support, from our F.D.Ls. to be made available to the snipers ; and where this is so it should be arranged.

28. When distance between the opposing sides is very great, and the enemy has no highly organized defence, snipers can be sent well out in front of our own lines. They will then operate on beats, rather than from static posts, with the object of stalking whatever enemy positions they can find. This will gain information about the enemy, as well as helping to further an offensive policy. It will sometimes be worth while to send out a small patrol, with a party of snipers. This will establish itself in no-man's-land, as a firm base from which the snipers can operate.

29. Careful consideration should be given as to the most suitable times for snipers to operate, as well as to the place from which they should do so. The limited number or snipers available makes it impossible to have them operating at all times, and quite frequently the best chance of their getting a shot will be in the half-light of dusk and dawn, or during moonlight. This is particularly so where our superiority of fire has driven the enemy to ground, during the hours of daylight, or when there is much enemy patrol activity by night.

Part II

Introduction

30. This part is for the guidance of officers and N.C.Os. responsible for the training of snipers. It is divided into three chapters, on observation, fieldcraft and rifle shooting respectively, which have been made as comprehensive as possible, so that courses of training can be based on it.

Sniping is, however, nothing more than the basic task of the infantry soldier, developed to such a degree that it has become a specialist one. There must, therefore, be a certain amount of overlapping between this pamphlet and certain other textbooks of basic infantry training. Moreover, there are certain subjects, mentioned as being useful to a sniper, that are fully dealt with in other pamphlets, and cannot be included here.

Thus the sniping instructor must possess a thorough knowledge of the relevant matter in Small Arms Training, Volume I, and Infantry Training, Part VIII. For map reading and compass work and for panoramic sketching, Notes on Map Reading must be consulted. Reporting and Photo Reading are dealt with in the Manual of Military Intelligence, and information about enemy equipment and identifications is given in various intelligence publications.

Chapter 2—Observation

Section 8   General

31. A highly developed sense of sight is essential to a sniper, not only because it is on this that he depends to find his quarry, but also because it will afford him protection against his enemies. Besides this he may often be called upon to help in the task of collecting intelligence, and

it is fitting that he should excel in what is an integral part of his specialist training. Observation is the first step towards offensive sniping, and in this, as in the subject of fieldcraft, it is necessary to re-develop a natural sense, which has been allowed to deteriorate, by conditions of security and civilization. The shikaris of India, the trappers of Canada, the hunters of the European forests, deer-stalkers and poachers, and all who have to pit themselves against wild animals, have retained the quick and perceptive eyesight that becomes of great value in war. But the townsman whose eyes need seldom exert themselves for day-to-day existence, can only develop a keenness of vision with much training and practice.

32. Training of the eye.—Proficiency as an observer comes rather more from a mental attitude than from physical endowment, and both of these qualities must be considered.

As regards the physical aspect, although no two men may have equally efficient eyesight, in most it can be developed to an adequate degree. This involves purely and simply a muscular conditioning of the eyes, through practice, like marching conditions the body for route marching, or boxing improves the muscles for boxing. And so, as for the sprinter and the marathon runner, each of whom requires to train for a special type of running, the observer should train his eye over the distances, and in the outdoor surroundings that the circumstances of war will demand.

33. Mental alertness.—In the mental aspect, even more than in the physical, the observer can learn from the hunter, by emulating the hunter's alertness and patience, his attention to detail and powers of deduction, which must be of the highest order, to lead him to his quarry. A hunter learns to notice scores of little things, the bending of a blade of grass, the unnatural shape of a shadow, or a disturbance of the undergrowth, from which he can deduce causes that would escape the casual observer.

34. It is from such signs that a good observer will often learn most about the enemy, for when the enemy is skilful in concealing himself, his actual troop movements will be seen very rarely. At such times much can still be learnt of his whereabouts and activities. An open window, that was closed before, a wisp of smoke, or a fowl disappearing from a derelict farm yard, all carry their significance.

The actions of birds and animals, differing according to whether they are wild or domestic, frequently give a clue as to the presence of man ; and a close study of the ground under observation will often draw attention to any future change in its features. Such changes may reveal the construction of a new enemy post, even though the post itself is cunningly camouflaged.

35. Mental alertness and the ability to notice such details as these with the faculty of drawing conclusions from them, is, therefore, of great importance to the sniper. The good observer must also possess patience, and thoroughness. If he notices an object of suspicion, the nature of which is not immediately obvious, he will mentally examine it from every angle, and will not rest until he has made it out. For such close scrutiny the telescope is an unequalled aid.

36. Light.—The sniper must know that the light is always changing and that almost from minute to minute, the visibility of objects will change with it. For him this means that his watching must be constant, so that no opportunity is missed of making some new find. He must also know when the light is likely to be in his favour and when otherwise. If the sun is behind him he will have the advantage ; and when it is low to his front, observation will be difficult, and the lenses of his glass may flash and give him away. The clearest light often comes just before or after a rain shower, and he must be quick to take advantage of such opportunities.

37. Use of glasses.—The use of glasses takes a high place in his training. The binoculars and observer's telescope, with which the sniper is equipped, are indispensable to him, and he must master their use. Binoculars are comparatively easy to use, and are valuable both in assisting a quick survey of the ground and also for observation by night. But the use of the telescope requires much training and practice, if good results are to be obtained. It must be brought home to the sniper, by practical demonstration if possible, how much detail the telescope will reveal ; for its high magnification makes it greatly superior to binoculars, in picking up small or well-concealed targets, and also for observing at long ranges. In both these respects the telescope is unsurpassed, which is why the deer-stalkers of Scotland carry it in preference to a pair of binoculars.

38. System in observing.—Whether or not he is using glasses, the observer must cover systematically the whole of the area for which he is responsible. Unless he does this he is bound to miss many items of interest. According to the nature of the ground, the following systems, or combination of them, will be found useful.

(a) Division of the ground into sectors, as dictated by the natural lines of roads, hedgerows, etc. Each sector can then be searched separately.

(b) Division into foreground, middle distance and distance. These, however, are not always easy to define.

(c) Working over the ground, or a sector of it, in strips. This method is well suited to the use of glasses, and the strips should overlap slightly, in the manner of a mowing machine mowing a lawn. The instrument should be moved slowly, so as to allow each field of view to be examined, before passing on to the next.

39. It is a natural inclination to pay particular regard to what appear to be likely spots. This is, in fact, sound, so long as the rest of the area is not neglected. It is a part of the skill of the observer to know what, in fact, are likely places to find the enemy. But he must also remember that the enemy, for his part, will try to avoid the obvious.

40. Familiarity of objects. Some effort must be made to make the sniper familiar with the appearance of the kind of objects for which he will be searching. If this is done he will stand a far better chance of recognizing them under difficult conditions. Familiar objects will always tend to arrest the eye ; but unfamiliar ones, or even familiar ones seen from an unusual angle, or draped with camouflage, may easily pass for an item natural to the countryside. A study should, therefore, be made of enemy weapons and camouflage ; and if possible these should be used in training.

41. Fatigue.—Comfort, combined, of course, with concealment, is very necessary towards getting good results. An observer must be prepared to keep patient watch for long periods on the chance of seeing even one item of note, and if he is uncomfortable he will be unlikely to do so.

Observation, moreover, entails great eyestrain, and it is necessary to cultivate unstrained watching to minimize this. A telescope that is slightly out of focus, or an uncomfortable position, will aggravate strain ; but it can be avoided to some extent by using alternate eyes, and, if possible, taking turns with another man. Do not observe with one eye screwed up tight, even with a telescope ; and look at something green, such as grass, to rest tired eyes.

42. **Reporting.**—Snipers should be trained always to keep a log of their observations, so that the value of these will not be lost to Intelligence. Such a log, a specimen of which is given below, should always be handed in to the Battalion Intelligence Officer at the end of a sniper's tour of duty. Some training should, therefore, be given to snipers in reporting. Accuracy in any of their reports must be made a point of honour with them ; but their deductions and suspicions are also valuable, and should be encouraged, so long as they state them specifically as such. A simple panoramic sketch is always a most useful accompaniment to a report, and some training should also be given in how to make them.

OBSERVERS' LOG

Names of Observers: A. White, Cpl. B. Black, Pte. *Date:* 12. *Tour of duty:* 0800-1830.
Position: No. 2 Post 373641. *Visibility:* Moderate.

Serial	Time	Map Ref.	Object seen	Remarks or action taken
1	0905	384655	Bunker at X Roads seen to be unoccupied.	
2	0930-0936	389661	Own artillery shelling road.	30 rounds fired on target.
3	1020	379682?	Enemy wire observed south of BIG WOOD.	Triple concertina bearing 11° mag.
4	1415	388649	Enemy O.P. suspected in Red Gable end.	Reported to I.O by telephone.
5	1505	382647	Two enemy observers spotted in hedge.	Range 300. Fired two shots at 1615 hrs. One hit claimed.
6	1620-1625	378649	Own artillery shelling Red Gable end.	Three direct hits. Two enemy ran out. Both wounded.
7	1710	377647	Enemy M.G. Post located under large bush.	

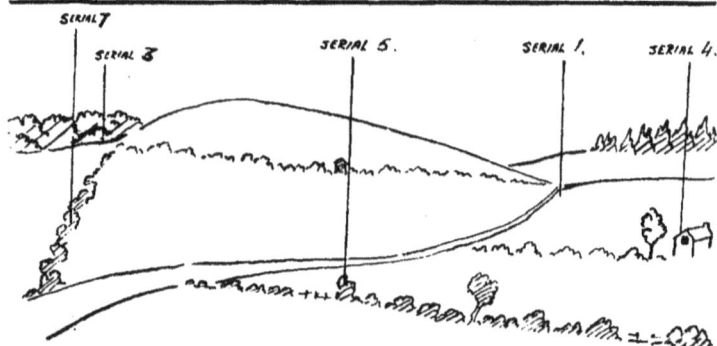

Sketch of View from No: 2 Post.
(to accompany log dated 12.)

43. Night observation.—Training is also necessary in observation by night. This differs from daylight observation mainly in that the sense of hearing becomes of greater importance in proportion as the value of sight decreases. The training can, therefore, be divided into two parts—visual training and training in hearing.

44. Visual training.—Night observation differs from observation by day, not only because there is less light to see with, but also because the eye uses a completely different set of nerves in the dark. By day, or in artificial lighting, these night nerves are blinded and are not used at all. By night the day nerves are useless. Night sight varies in different men, but in all cases it can be greatly improved by practice. The most important considerations affecting visual training by night are as follows :—

(a) The night nerves are more numerous towards the sides of the eye than in the centre. The most detail is, therefore, seen by looking slightly away from an object rather than directly at it.

(b) At night the eye " sees slowly ", and it may be necessary to look for several seconds in one direction before an object becomes visible. When searching, therefore, the eye should move from point to point, resting for a few seconds at each point, and again remembering that the keenest vision is obtained away from the centre of the field of view.

(c) Broad shapes and silhouettes only will be seen, small details being invisible by night. Objects will, therefore, tend to look different from what they do by day. For example, trees seem smaller by night, because the twigs are not visible. This makes it very difficult to judge distances, and also makes a daylight reconnaissance very desirable.

(d) The night nerves are colour-blind. An object is only visible from its contrast in tone, with that of its background. Light objects show up against dark backgrounds, and *vice versa*. They show up best of all against the sky ; so the observer should keep to the low ground, to obtain an extensive skyline.

(e) By night, moving objects are much more easily seen than stationary ones.

(f) The eyes will tire very quickly at night, though the observer himself will probably fail to notice when this is so. They should, therefore, be rested for 10 seconds, after every two minutes of observing.

(g) Unstrained searching must be cultivated. It is important not to stare at objects, or the eye will quickly tire, and the object itself will appear to move ; so that inanimate objects such as fence posts or bushes may be mistaken for men. This is another reason for careful daylight reconnaissance.

(h) The night nerves take up to 30 minutes to gain full efficiency, after being in a bright light. This should be remembered when going out into the darkness from artificial lighting. The accommodation of the eye can be lost in a few seconds, by looking at a bright light, which must, therefore, be avoided. Where this is unavoidable, the ill-effects can be minimized by closing one eye. When reading a map with a torch this should, in addition, be shaded as much as possible.

(*i*) Binoculars will give good results on all but the darkest nights, and should always be carried. It is important that they should be correctly focused or a serious loss in efficiency will result. Care is needed over this, because incorrect focusing may not be noticed by the observer, even though it is present. The same applies to the correct setting of the distance between the eye-pieces ; and dirty lenses must also be guarded against. Night focus will differ slightly from the normal day focus, and can best be adjusted on some object, such as a tree, silhouetted against the sky.

45. Hearing. — The sniper should be practised in recognizing sounds by night, and in locating their direction and distance. Sounds carry farther and more clearly in the dark than they do by day, which makes their distance extremely deceptive. Their strength is also influenced by weather conditions and the nature of the country. It is a good thing to find out which is the master ear, because this will influence the sense of direction of sounds. The hearing is most acute when the head is uncovered, and for picking up some slight sound is further assisted by closing the eyes and slightly opening the mouth.

Section 9   The telescope and binoculars

46. The telescope, scout regiment

(*a*) *General.* A telescope is a delicate instrument, and can easily be damaged unless properly cared for. The model authorized for snipers is the Scout Regiment, Mark 2, which has a magnification of 20. Other models may be met with on service, but all have similar characteristics. The high magnification of the telescope makes it possible to pick up small objects at great ranges, but its small field of view necessitates skilful handling. It is an essential part of the sniper's equipment.

(*b*) *Description*

    (i) The telescope consists of three gioups of lenses, contained in four metal tubes, which slide into one another.

    (ii) The largest of the four tubes is called the body, and contains the object glass cell. The front part of the body slides out over the object glass, and is called the rayshade, and the leather covering of the body is called the handguard.

    (iii) The smaller sliding tubes are called draw tubes, and are attached to one another by means of cells, which contain packing made of felt or some similar material.

    (iv) The object glass consists of two lenses held together in a cell, which screws into the front of the body. In some telescopes the two lenses are stuck together by a thin film of optical cement.

    (v) In the front end of the rear draw tube is the erector cell which holds two lenses, the field lens in front and the eye lens in rear. Without this group, everything seen through the telescope would appear upside down. In the middle of this cell is a diaphragm, with a small hole in the centre called the pupil.

(vi) The eyepiece cell, which also contains a field and eye lens, slides into the rear end of the telescope.

(vii) Over the rear of the eyepiece cell is screwed the eyepiece itself, which contains a small shutter to exclude dust from the lenses. In some telescopes this shutter contains a window of tinted glass, to assist observation in a bright light.

(viii) The telescope is focused by fully extending all the draw tubes and then pushing the rear draw tube slowly forward, until a clear view is obtained. This should be done with the eye gazing through the telescope in an unstrained manner, otherwise the eye may adjust itself to the telescope when this is not in perfect focus, and eyestrain will then inevitably result. If both eyes can be kept open during the operation this danger will be overcome.

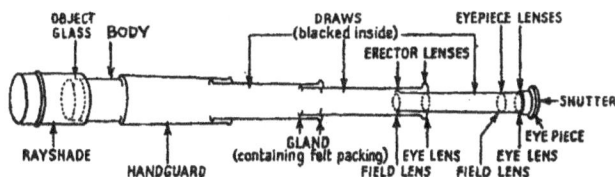

Telescope

(c) *Stripping*. The telescope should never be stripped except for cleaning purposes, when dust or damp have entered into the draws. Excessive stripping will cause wear to the threads of the screws. When re-assembling any of the screw-in parts, these should first be turned in an anti-clockwise direction, until there is a slight click, and then screwed up gently. Force must never be applied or cross-threading and stripping of the threads will result.

(d) *Cleaning*

(i) The lenses are of very soft glass, highly polished and easily scratched by hard or dirty cloth. Grease from the fingers or rainwater, if left on, will permanently mark them. They should be polished lightly, when necessary, with clean cambric or flannelette. A piece of this should be kept for the purpose in the top of the telescope case, or in the pay-book. The lenses must never be rubbed hard, or they will become marked. If grease gets on them, it can be removed with methylated spirit.

(ii) If the telescope gets wet, the lenses will fog up on the inside and the packing will stick. Before returning it to its case all water should, therefore, be wiped off the outside of it, and later it should be extended in some dry place, with the object glass and eyepiece cell removed. It must not, however, be placed close to a fire, particularly if the object glass contains Canada Balsam, since this melts at low temperature. Melted Canada Balsam makes the lens opaque.

(iii) If the draws or the rayshade are hard to open as a result of getting wet, after being throughly dried they may be lubricated with powdered chalk or a very thin film of light oil. Oil must be used very sparingly and with care, to prevent any of it getting on to the lenses.

(e) *Care of the telescope*

DO (i) Always draw the rayshade first. Many O.Ps. have been given away by the sun flashing on the object glass.

(ii) Draw and close the telescope gently, and with a clockwise turning movement. This will prevent damage to, or unscrewing, of the glands.

(iii) Mark the focus with a ring scratched on the rear draw, to facilitate quick focusing.

(iv) After closing the telescope, shut the eyepiece shutter to exclude dust.

DON'T (i) Touch the lenses with the fingers. The skin always contains a certain amount of oil.

(ii) Bang or jolt the telescope. The draws will dent very easily, and the blacking on their inside will chip off and get on to the lenses.

(f) *Handling of the telescope*

(i) Correct holding is vitally important. To get good results the telescope must be held still, and this is impossible unless it is held comfortably. It must, therefore, be rested in some way.

(ii) The methods of holding are :—

Lying on the side, front hand holding handguard, and rested on the lower portion of the thigh. Rear hand holding eyepiece between first and second fingers, with thumb rested against the nose.

Standing or sitting with a rifle or stick to support front hand.

From behind cover, with front of telescope rested. A short forked stick will assist in supporting the front of the telescope.

Round the side of cover with handguard rested against the cover.

Sitting back to back with another man. Elbows rested on knee.

(g) *General hints in using the telescope*

(i) To pick up a particular object, take a sight along the top of the telescope. Then drop the eye to the eyepiece, without moving the telescope.

(ii) Use it with both eyes open, because this greatly assists in avoiding eyestrain. This may be found difficult at fi but must be practised. It can be learnt by focusing telescope on to some object with the other eye shut. concentrating on the object through the glass, and t allowing the other eye to open, the view through the g will be retained, and the other eye will " idle."

(iii) Allow fields of view to overlap, and search each thoroug.

(iv) To memorize the position of an object found, note its position in relation to two or more landmarks, and look away for a few seconds. Then look back, and pick it up again.

(c) Sub-para. (g)(ii). Add at end of paragraph, after "other eye will 'idle'", "With one eye shut, the observer will tire after ten minutes or so. With both eyes open, observation can be continued, even in a bright tropical light, for several hours on end without any undue strain".

Handling the telescope

Lying on the side.

Sitting, with rifle
used as support.

Lying behind cover using forked stick.

47. Binoculars

(a) Binoculars generally have a magnification of between 6 and 8 diameters. This, with a wide field of view, and general handiness, make them quicker to use than the telescope, though they will not reveal many of the targets visible with a telescope. They are also of great value for night observation, and, as an adjunct to the telescope, form an essential part of the sniper's equipment.

(b) The principles regarding their care and cleaning are similar to those applying to the telescope. They must, however, never be stripped. The combination of lenses and prisms, which they contain, have to be most exactly set by skilled opticians, and the slightest disturbance of them will render the instrument useless. Every effort is made to make the No. 2 binocular as far as possible waterproof, but even so the instrument will "breathe" when changes of temperature occur, and moisture in the atmosphere, sufficient to fog the lenses may enter. This "breathing" must be prevented, as far as possible, by screwing the eyepieces home when the binoculars are not in use.

(c) Focusing is effected either by an adjusting screw, incorporated in the central bracket, or by adjustable eyepieces.
In the former type, one of the eyepieces is also adjustable, to allow for differences in the eyes of the user. To obtain correct focus, first focus the set eyepiece, keeping the other eye closed. Then, with both eyes open, and looking through the binoculars, focus the adjustable eyepiece.

Section 10   Location of fire

48. General.—The location of enemy snipers or machine gun positions will often be one of the sniper's most difficult tasks. Provided that certain principles are understood, much information can, however, often be obtained from the sounds of the enemy fire itself.

When a bullet is fired, having a velocity greater than that of sound, the bullet itself will make a loud noise in forcing its way through the air. This noise is quite separate from the report made by the weapon which fires it. Therefore, if a bullet is fired from a modern rifle or machine gun, towards the observer, the first noise that he will hear is a sharp crack, made by the bullet as it passes by. Later, and generally less distinct than this crack, comes the duller thump of the explosion.

49. Crack and thump.—When the range is short the time lag between the crack and the thump will be very slight, but as the range increases it will become longer and larger. There comes a point at a range of about 1,000 yards, when the average service bullet will have slowed down so much that the time lag will start decreasing. But below this range it is possible, with practice, to judge the approximate range of the firer by this time lag alone.

The natural tendency of the observer will be to look in the direction of the bullet's crack. But this must be counteracted, for it is the thump of the explosion that will give a clue as to the firer's direction.

50. Procedure.—The sniper must, therefore, be given practice in hearing bullets fired in his direction from different ranges, preferably from enemy weapons. With experience, he will then become expert at judging their range and direction. The following is a useful procedure to follow in locating the enemy by this means :—

(a) Judge the range to the firer, from the interval between crack and thump.

(b) Judge his direction by listening for the thump.

(c) In the area so determined, watch for the muzzle smoke of the weapon.

(d) If this cannot be seen, then search the area, to which his position has been narrowed down, with the telescope or binoculars.

Section 11   Notes on observation training

51. Frequent practice should be given to the sniper in observation with the naked eye, with binoculars, and with the telescope. Only by constant practice will efficiency be obtained in handling the binoculars and more particularly the telescope.

Correct handling of both these instruments and system in searching ground must be insisted on until they have become second nature.

Observation schemes must be progressive, and some suitable ones for initial training are outlined below. Enemy uniforms and equipment should be used when available.

When these initial exercises have been completed, more difficult variations of them can be devised, and opportunities taken of observing troops on training. Any spare ten minutes can be usefully filled by practising the handling of the telescope and quick focusing on moving objects, such as passing vehicles.

Night observation schemes with binoculars should also be included in the sniper's training.

52. **Initial observation scheme5** .

(a) *Object.*—To pick out partially concealed inanimate objects, *e.g.*, steel helmet, pick, shovel, pouches, rifle, bayonet, boot, etc.

Range.—100–300 yards.

Stores required.—Telescopes notebooks, pencils, groundsheets.

Notes. — Men to work in pairs, one observing, one keeping log, then change round. Arc to watch is given. What seen and where and estimated range to be entered in log.

Class observes with naked eye—5 minutes.

Class observes with telescopes—20 minutes.

Collection of logs and objects, and discussion.

(b) *Object.*—To detect men using natural cover.

Notes.—Ten men concealed with rifles. Remainder Scheme (a)

Object.—To detect men using camouflage.

Notes.—Ten men concealed, using camouflage devices—100 to 300 yards. Remainder as for Scheme (b).

Object.—Detection of long range movement.

Notes.—Fatiguemen simulating movement of enemy patrols and working parties from 200 to 2,000 yards.

Numerous varieties of this form of exercise are possible, and they can be combined with location of fire exercises. It is also possible to make successful observation of troop movements up to 10 miles or more, in a good light, and observers should be given some practice in such long range work.

Chapter 3—Fieldcraft

Section 12   General

53. Fieldcraft is the use of natural cover, when stationary and during movement, permitting the free use of arms. It is a part of the hunter's art, and consists of the silent movement and use of cover, the knowledge of his prey and skill at arms which enable the hunter to make his kill. Fieldcraft, therefore, demands physical fitness, mental alertness, patience and discipline. It also calls for the ability to read and use ground.

54. These attributes, when developed and applied with initiative, make a man a formidable fighting opponent. The sniper must, therefore, develop the same cunning that the hunter uses to outwit his quarry, and since his quarry is another man, must know how to outwit him when he tries to retaliate.

55. Knowledge must be acquired, not only of the concealment afforded by vegetation and other features covering the earth's surface. Every fold in the ground, however slight, may afford cover, and to know where such folds exist, and how to use them, requires much practice. The protection and dangers afforded by light and shadow must also be studied, and in movement both agility and stealth must be cultivated. Fieldcraft training should aim at developing a constant alertness of the mind and the senses of sight, hearing, smell and touch.

Section 13   Concealment

56. When stationary, the problem of concealment calls for a knowledge of how best to use the natural cover which the ground affords, and of the assistance given by camouflage.

(a) *Use shadow. See para.* 57 (a).

57. Natural aids.—As regards the use of natural cover, the following points are fundamental and should be so mastered as to become instinctive :—

    (a) **Keep in shadow.** This is the greatest protection of all, but remember shadow moves with the sun, and do not break the natural shape of a shadow.

(b) Match your background.

    (b) **Background** which matches the colour of the clothing can be as good protection as cover in front of you.

(c i) Avoid Skylines—there are two men here.

    (c) **Avoid breaking skylines and natural straight lines.**

3a—22811

(c ii) *Don't break straight lines—there are two men here.*

(d) *Don't look over cover. Look round or through it !*

(d) **Look round the side of cover, rather than over it. Look through it,
where this is possible.**

(e) Avoid unnecessary movement. Nothing catches the eye so quickly.

Therefore, merge into your cover and fade out of it.

Go HERE

NOT here

(f) *Avoid isolated or conspicuous cover.*

(f) Avoid isolated or conspicuous cover to which the enemy is likely to pay particular attention.

Wrong!
The face
and hands
will shine

Blacken
them

Or use
a face
veil

Or hold a
spray of leaves
in front of
the face

(g) The face and hands, unless camouflaged or concealed behind cover, are outstandingly obvious.

58. These points should be demonstrated to the sniper at the outset of his training by the use of fatiguemen.

The demonstration could take the following form :—

(a) Opening remarks	5 mins.
(b) Squad search ground for men in wrong positions ...	3 mins.
(c) Point out that nearby to each of the men is cover, which, if properly used, would give much better protection. Enumerate faults from right to left ...	5 mins.
(d) Turn squad about. Signal men to take up correct positions	2 mins.
(e) Squad face front, search ground	5 mins.
(f) Signal each man from right to left to stand up and expose himself, pointing out why each obtains better concealment than before	10 mins.
(g) Recapitulate points brought out	5 mins.

59. **Camouflage.**—Concealment is further aided by the use of camouflage. Some animals are provided by nature with wonderfully effective camouflage, but the soldier has to make it for himself. The aim of personal camouflage is to hide the familiar shape of the man, and the shine from his clothing, skin and equipment, while matching him, as far as possible, with his surroundings. The interests of concealment and camouflage, however, conflict with those of mobility and weapon handling, so that some degree of compromise is always necessary. Shine can only be prevented by making surfaces rough, and shape is distorted by loose-fitting garments, both of which can be a hindrance, if carried too far.

Unless some special colour scheme is used to match a particular background, and this is seldom possible, this also calls for compromise ; and, for general use, a combination of neutral colours, greys, greens and browns, is best. But some contrast both of colour and tone must always be introduced, to obtain a disruptive effect, and small amounts of highly contrasting colours, such as yellow or black, will be a help, if used sparingly.

60. Normally, the only practical measures that a sniper can take towards personal camouflage, are on the following lines :—

(a) *Head.*—Garnished helmet net, or cap comforter worn on the head. Face darkened and further concealed, where necessary, with a face veil, which should also be garnished.

(b) *Hands.*—Darkened with sufficient camouflage cream or covered with dark gloves.

(c) *Body.*—The Denison smock, an issue to snipers, will normally be worn. It can be improved for purposes of concealment by stitching on a few tufts of hessian garnish. Other types of suit for snipers may be forthcoming in the future.

(d) *Rifle.*—Some protection may be gained by disruptively painting the barrel and fore-end. Shiny metal portions should also be treated in this way.

(e) *Telescope.*—When in use, the body and rayshade of the telescope should be covered with hessian or an old sock with the foot cut off.

(f) *Binoculars.*—A strip of hessian can be tied round the front of the bracket. This must not interfere with the focusing gear.

61. The sniper will seldom have occasion to employ any more complicated devices than the above. He should, however, be given an opportunity of studying the effect of other forms of camouflage, whether issued or improvised, since these may, on occasions, be useful to him. This will also help him to see through the enemy's attempts at camouflage, and forms excellent practice in the study of the protection afforded by various types of background.

62. A demonstration should be staged on the following lines :—

(a) Opening remarks ... ... 5 mins.

(b) Squad searches ground with naked eye for camouflaged men, using a wide assortment of camouflage, against correct background ... ... 5 mins.

(c) Turn squad about. Signal men to take off camouflage, and adopt the same positions as before ... 3 mins.

(d) Squad searches ground and instructor points out that each man is in the same place as before ... ... 5 mins.

(e) Signal men to wave a white flag, if any not seen, and tell squad to note the position of each man ... ... 2 mins.

(f) Turn squad about and signal men to adopt camouflage again ... ... 3 mins.

(g) Squad looks at each man with glasses and notes points that make him visible. Instructor points out that correct use of camouflage and background will give the degree of protection shown. This requires an eye for background ... ... 10 mins.

(h) Fatiguemen are called in, and camouflage devices inspected. Instructor points out which are issued and which improvised ... ... 7 mins.

Section 14   Movement

63. **General.**—The sniper must be well practised in the movements of a stalker, so that he can employ them efficiently and without fatigue. These movements include a number of different methods of crawling to suit various types of cover, and the ability to walk silently. When carrying them out, the sniper must be able to use his weapons instantly, if need be, and to maintain an all-round watch.

64. The most useful movements to practise are as follows :—

(a) *Walking silently.*—Care must be taken in placing the feet on the ground. When crossing hard ground, the least noise is made if the outside of the sole of the foot is placed first on the ground. Balance is most essential to silent movement, and this is assisted by keeping the knees slightly bent, and the arms hanging loosely downwards ; from this position it is possible to drop quickly to cover. It is a mistake to walk with the head bent down, because this affords no added protection, and impedes observation.

(b) *Crawling on hands and knees.*—This is useful for cover over 2 feet high. It can be carried out quickly with practice. But this is not always desirable, and for silent movement the knee should be placed in the very spot vacated by the corresponding hand. The hands should reach forward for a safe place to avoid cracking twigs and other obstructions.

Cautions.—Keep the head and posterior down.

Walking—carriage of arms.

Alternative method

Crawling
on hands
and knees

without
arms,

with rifle.

(c) *Crawling with the elbows and knees.*—This is useful for lower cover. The body and head must be kept close to the ground, propulsion coming from alternate elbows and knees, accompanied by a slight rolling movement of the body as each knee is flexed. As an alternative, one knee alone can be used, the other leg trailing flat along the ground.

Cautions.—Avoid kicking up the heels. Keep the head, body and elbows close to the ground.

Crawling with elbows and knees.   **Method of propulsion.**

Alternative methods

of carrying the rifle.

(d) *The stomach crawl.*—Where extreme caution is needed, or the cover is very low, the whole body must be pressed flat to the ground. The body is dragged forward by the elbows and progress is slow and tedious, but in this way very bare stretches of ground can be crossed without being observed.

Cautions.—Heels must be kept quite flat and the face almost touching the ground.

(e) The sniper must also be practised in backing away from cover, and in turning round in the prone position, particular care being necessary not to expose the heels, which must be kept flat.

The stomach crawl.   **Method of propulsion**

without arms

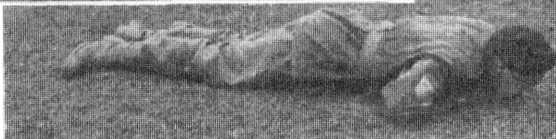

and with rifle.

Turning round in the prone position.

Three movement stages in making a 180-degree turn.

65. Carriage of arms.—The carriage of arms throughout these movements requires attention. The rifle must be so carried as :—

(a) Not to reveal the movements of the sniper.

(b) Not to impede his movements.

(c) Not to get dirt in the sights, barrel or mechanism.

(d) To be available for instant use.

66. As may be seen from the preceding photographs, the rifle is carried to comply with those requirements in the following ways :—

(a) In any movement, held with the left hand at the point of balance, muzzle pointing slightly upwards and to the left front. Butt slanting across the body to the right. Sling caught up in the left hand to prevent it getting in the way.

(b) When walking, with the left hand at the point of balance and the right hand holding the small of the butt. Muzzle pointing downwards and to the left front. In this way the rifle can be used more quickly than in the previous method, but the previous method leaves the right hand free.

(c) When crawling on the stomach or on the elbows and knees, by holding the sling, with the clenched hand, where it joins the outer band, muzzle to the front and the rifle resting along the forearm.

67. In practising all these movements, the sniper must learn to keep constant watch to his front, flanks and rear, yet without raising his head one inch more than is necessary to do so. He must learn when to move quickly and when to go at the pace of a snail, when to freeze and when to forestall the unexpected with a quick shot. When cover is scarce, slow movement and a favourable background may prevent detection. When crossing short gaps in cover, a quick dash may be the best policy.

Section 15   Stalking

68. **General.**—Stalking is the application of fieldcraft in its widest sense to bring the sniper within range of his quarry. It cannot be taught in the matter of a few lessons, but can only be learnt by experience. It is possible, however, to give some general principles which will always be of assistance.

69. **Reconnaissance.**—Thorough reconnaissance of the ground is essential, before starting out on any stalk, and careful observation of the following points will invariably be repaid :—

(a) The exact location of the enemy to be stalked must be memorized with reference to nearby landmarks.

(b) The final objective or fire position to be gained must be chosen, and similarly memorized.

(c) The whereabouts of any other enemy posts must be spied out. For this the stalking telescope will generally be required.

(d) The route must be decided on by weighing up the advantages and disadvantages of the ground. Considerations affecting the choice of route are :—

    (i) Cover and dead ground.

    (ii) Obstacles, natural and artificial.

    (iii) Points of observation en route, *i.e.*, bounds.

    (iv) Other known or probable enemy positions.

70. **Keeping direction.**—Great attention must be paid to keeping direction. This is not at all easy, especially when crawling or when using the cover of darkness or fog.

(a) The use of the compass and map, and of air photographs, must have been mastered, though the latter may not always be available.

(b) A distant landmark, or better still two, can be memorized.

(c) A series of landmarks along the route can be memorized.

(d) The direction of the wind can help in keeping direction.

(e) So can the position of the sun.

(f) The route back must be remembered ; and this can be done by looking back frequently on the way out.

71. Points to note.—The following points should also be remembered when actually engaged on a stalk :—

 (a) Alertness must never be relaxed for a moment.

 (b) Observe carefully from bound to bound.

 (c) Have an alternative route ready in case of meeting the unforeseen.

 (d) If surprised, or exposed, either freeze like a stone, or move quickly and get clean away from the point of exposure.

 (e) Think out ways of drawing the enemy's attention away from you. This may require somebody else's co-operation, or you may make something for him to look at in the early stages of your stalk.

 (f) Avoid disturbing animals and birds.

 (g) Take advantage of other movements and noises going on, to distract attention from your own.

 (h) Do not crawl where walking is possible.

 (i) If risks have to be taken it is better to take them early on.

72. Stalking by night.—The sniper will often have to move by night either because his fire position may not be reached by daylight, or when making use of his telescopic sight for night firing. Night stalking presents the same problems as stalking by day with the additional one that man is not a nocturnal animal. The main differences between night and day which affect the sniper are that by night :—

 (a) Protection is given against aimed fire by the enemy.

 (b) Sight is largely replaced by hearing, so that silence is of prime importance.

 (c) Background becomes relatively more important than cover, and avoidance of skylines is of increased importance.

 (d) Keeping direction is of even greater difficulty, and daylight reconnaissance of the route to be taken and landmarks along it is more necessary. A knowledge of the stars will be of great assistance in keeping direction.

73. Much practice is required in moving silently by night. Precautions must be taken to prevent equipment from rattling, and parts that are likely to rattle should be muffled with strips of sandbag material or adhesive tape. Particular care is needed in placing the feet on the ground and in preventing them from making a noise in long cover or on stony ground. Silent movement in the dark depends largely on the sense of touch. Frequent halts are necessary to observe and listen. Anything that may shine, including the face and hands, must be obscured and luminous watches or compasses may be a source of danger. Practice is also required in crossing obstacles, both natural and artificial, in the dark.

Section 16   Sniping from hides and buildings

74. General.—During static periods and in defence, snipers should, whenever possible, build hides, because more efficient work can be done when :—

(*a*) A certain amount of free movement without fear of detection, and

(*b*) Some protection from the weather is possible.

Hides can vary considerably as to type ; and it is not possible to list them all. The type will depend on ingenuity and the ability to improvise.

Some of the many possible locations are under hedgerows, in ruined buildings, a disabled tank, a rubbish heap, the edge of a cutting or a stream bed.

They may be sturdily built of stone, brick or wood, but more often only the scantiest building materials, to give cover from view and the weather, can be used. An enlarged slit trench with some improvised head cover, will often be the best answer. It is also desirable, but not always possible, for them to be bullet-proof.

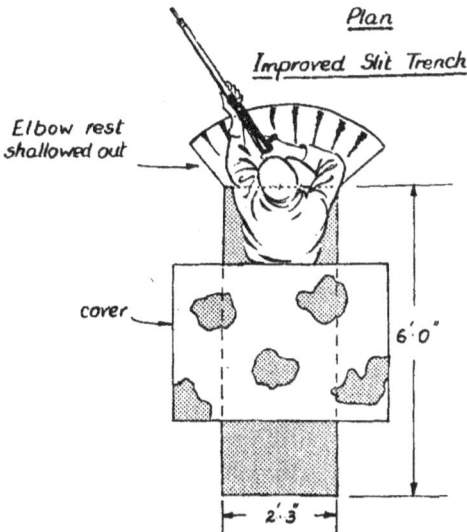

Plan

Improved Slit Trench

Elbow rest shallowed out

cover

6′0″

2′3″

75. Concealment

(*a*) *Camouflage must be carefully arranged.*—It is essential that this should be of the highest order, and that the natural appearance of the ground should not be changed. It should always be borne in mind that cover from view is cover from aimed fire, as long as the sniper is not seen getting there. Spoil must be hidden. Conspicuous landmarks must be avoided.

(*b*) Movement in and around the hide must be avoided. No one should visit it except the snipers. Track discipline must be observed and unless there is a good covered approach, the hide can only be entered and left during darkness.

(*c*) Unnecessary movement inside the hide must be avoided. Practice is needed in the art of keeping perfectly still for long periods.

(*d*) Firing from the hide at dawn and dusk will give away the position. Under cover of darkness, it is possible to use other positions away from the hide.

(e) Light shining through loopholes from the rear gives them away. There must be a screen over the entrance to the hide to prevent this, and the screen must not be moved unless the loopholes themselves have been screened or bunged up.

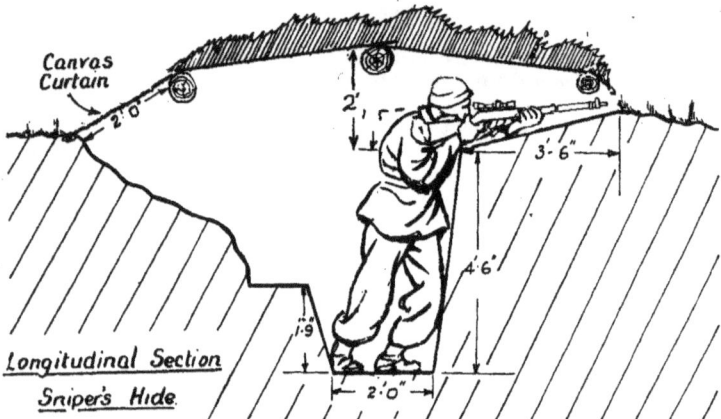

Canvas Curtain

Longitudinal Section
Sniper's Hide.

(f) The loophole and the ground outside must be well damped, if there is any danger of dust rising when the shot is fired.

(g) On frosty mornings, and damp days, smoke from the rifle shows most. When possible, the sniper should keep well back from the loophole to prevent this danger.

76. Comfort

(a) There must be room to use the rifle and to observe.

(b) There must be room for two men.

(c) There must be room for them to move their limbs.

(d) There must particularly be enough headroom to use the rifle with a telescopic sight.
These points appear obvious, but in nine cases out of ten, when a man constructs his first hide, he will have made it far too small, with no room for his head.

(e) A comfortable position to shoot and observe from, and, where necessary, a seat must be made.

77. Field of fire

(a) Care and practice are needed in constructing loopholes, so that they cover the required field of fire.

(b) The correct way of making a loophole is wide at the back, to allow an adequate traverse, and narrow in the front, but not too narrow to permit the use of the rifle and telescopic sight.

(c) Loopholes can be camouflaged by making them through old tins, old boots, or other rubbish that is natural to the surroundings. If the subject is not natural to the surroundings, it is a great source of danger. Natural camouflage can, however, often be most effective and shadow provides the best concealment of all.

(d) Since hides will almost always have to be made during darkness, the site should be reconnoitred whenever possible by day, and the direction for the loophole marked by two sticks or pick handles. Otherwise it will generally be found that the completed loophole is looking in the wrong direction.

78. Deception

(a) There should be as many alternative hides as possible and programmes of occupation should be irregular.

(b) Devices should be used to attract the enemy's attention away from the hide, but not on to neighbouring positions, e.g. dummy loopholes, or movements produced at the end of a length of string or telephone cable.

79. Buildings.—Buildings often offer the best opportunities as sniping posts, under static conditions. They have, however, the great disadvantage that they are bound to attract fire from the enemy's heavy weapons, and an isolated house is very likely to be shelled even if the sniper using it has not been detected.

Sniping from buildings

They should be prepared for use in much the same way as a hide. That is to say, similar precautions towards concealment must be taken, loopholes must be constructed and fire positions made. Care must be taken not to alter the outside appearance of the house such as by opening windows, doors or curtains, that were previously closed. The fire position must be well back in the shadow of the room, and window, or shell holes, against which the sniper might be silhouetted, must be screened.

Loopholes may consist of holes in window, shutters, or roof, or they may have to be picked out of the actual walls. In shape they should simulate shell damage, otherwise their front openings must be as small as is compatible with field of fire. Some form of rest must then be constructed behind them, to make accurate shooting possible and to assist in observation with the telescope. This and the fire position may be improved from furniture or else by sandbagging.

Section 17   Notes on fieldcraft training

80. **General.**—After being shown demonstrations on concealment, both with and without camouflage, the squad should be practised in taking up fire positions, under the observation of fellow-students and instructors. Later in their training this factor can be brought into field firing exercises, by having an instructor, with a periscope, at the target end, observing from bullet-proof cover.

Individual stalks over short distances, introducing the competitive spirit, give the best initial practice in movement and the use of ground. These must have been preceded by lessons in movement, and stalking in general, and also, where possible, by a demonstration stalk carried out by experts. Many variations of the specimen exercises, given below, may be devised and training can at the later stages be combined with map reading, observation and other factors. Finally, through field firing exercises, practice in marksmanship can be linked up with these other phases of training.

81. **The individual stalk**

    (*a*) The squad is divided into two halves. One half act as pairs of observers, and are placed in position, with an arc to watch. The other half are started from points about 200 yards away from one or other pair of observers, and have to approach to within grenade throwing distance of the observers, by independent routes.

    (*b*) The stalkers are given a few minutes to study the ground and select their objective and route. They are then allowed a limited time to carry out the stalk. The observers will keep a log of any exposures made by the stalkers, recording the length of the exposure and the reason for it. They will also have coloured flags, which they will raise, whenever any of the stalkers shows himself.

    (*c*) A discussion should take place at the end of the stalks, on points taught in the lessons on movement and stalking.

    Note.—This exercise, with slight modifications, is well fitted to practising movement by night.

82. **Infiltration stalk**

    (*a*) The squad is divided into pairs of stalkers, and an objective is pointed out 700 to 1,000 yards away. Between the stalkers and their objective are a number of enemy posts, manned by fatiguemen previously placed in position.

    (*b*) On a given signal the enemy open fire with blank ammunition, and continue to fire until a signal to stop is given. During this time the stalkers must locate the position of the enemy.

    (*c*) The pairs of stalkers then make their way to the objective, by independent routes, avoiding detection, if possible, by the enemy post. The enemy will fire further rounds of blank if they see any of the stalkers.

(d) A discussion will be held, at the end of the exercise, on the points of fieldcraft and observation which have arisen.

(e) Where conditions of ground allow and with care in arranging safety precautions, the enemy may use live ammunition. This will give added practice in location of fire, and greater realism. In this case the enemy must consist of reliable N.C.Os.

83. Sniping from hides

(a) Two parallel ridges, some hundreds of yards apart, form the most suitable ground for this exercise. A tactical setting is given, and one of the ridges is said to be held by the enemy. Snipers, working in pairs, are given a section of the enemy ridge for which to be responsible, and an area of the other ridge in which to build a hide.

(b) The area will be reconnoitred in daylight, under observation from the enemy ridge, and a hide constructed by night and manned before dawn. Small enemy patrols and other activity during the night help to maintain alertness.

(c) At dawn, movement is made on the enemy ridge, to represent the vacating of night posts, manning of O.Ps., etc., the snipers keeping a log of any movements they can observe.

(d) This exercise introduces the elements of patience and fatigue, and it is especially important to bring in a competitive element. If conditions will allow, the enemy can be withdrawn and the exercise can be concluded with field firing at figure targets, previously put in place.

Chapter 4—Shooting

Section 18   General

84. The sniper must become an expert in rifle shooting, and he must attain a standard of skill far higher than that normally demanded of the soldier.

For him the basic training, which he gets in his early days in the Army, is only a ground work, on which he must build a store of knowledge and experience, that will ultimately make him an expert shot. His special rifle is fitted with a telescopic sight, which greatly simplifies aiming at obscure targets and when the light is bad. But the telescopic sight will by no means make a mediocre shot into a good one, and the full benefits from it will not be obtained unless the sniper has first become a good shot without it. He must, in any event, be prepared to use the aperture sight of his rifle as well as the telescopic, since, under certain conditions, such as in rain, or if his telescopic sight becomes damaged, he will have to shoot without it.

85. The sniper rifle, using Mark 7 ammunition, is capable of grouping to within $2\frac{1}{2}$ inches or better at 100 yards, and the sniper's aim must be to shoot to within this standard. Further, he must aim at maintaining this standard proportionately at other ranges up to 600 yards. To do so he must not only be capable of holding consistently to such a grouping capacity under the ideal conditions of 100 yards firing point, but he must also understand the factors that will

affect his group, under the varying conditions met with at the longer ranges. Furthermore, he must learn how to apply his group to his mark ; and he must become a quick shot, able to hit a fleeting target in under three seconds if need be. In order to attain this standard, he must have an understanding of the following factors :—

(a) Sources of error in himself, and how to cure them.

(b) Sources of error in the rifle and ammunition.

(c) The effect of wind, light and weather.

(d) The adjustment of elevation and deflection.

(e) The accurate judging of distance.

86. This chapter deals with various aspects of the foregoing factors, with the exception of judging distance, which is fully dealt with in Small Arms Training, Vol. I, Pamphlet 2. The instructor should remember that, while the standard demanded is high, and the knowledge required by the sniper is extensive, any man who possesses normal eyesight and physique, can become a good shot ; but to make him one will require patience and individual attention on the part of his coach. One further thing is needed—keenness on the part of the man, and willingness to learn. Given this, a rifle can be made a deadly instrument in his hands.

Section 19

Sources of error in the man and the essentials of shooting

87. No man can fire a rifle without introducing some error through the human element. This error can, however, be cut down to something exceedingly small. Errors may arise in aiming the rifle, or in holding it on the aim, or through disturbing the hold by incorrect trigger release. These three essential factors, aiming, holding and trigger let off, are dealt with in Small Arms Training, Vol. I, Pamphlet 3, but too often the guidance given is indifferently applied. The sniper must first gain confidence in his ability to keep down his errors to some consistently low standard, which he will never unknowingly exceed. This, in other words, is the ability to group and to declare confidently when he fires a shot that will fall outside his normal grouping capacity. This is the bed-rock of all rifle shooting, and is absolutely fundamental. The main faults which occur at this stage are discussed below.

88. **Faults in holding.**—Wobble, or inability to hold the rifle still, may come from bad instruction, lack of practice and determination, or unfitness. Although it is never possible, even when using cover, to hold a rifle perfectly still, its movement may be cut down to something infinitesimal. This is very difficult unless comfort is obtained in the firing position, and thus it is most important that the butt of the rifle should first of all fit the firer, and that he should lie in such a way that the rifle will come up naturally pointing at, or very near, the target without strain. A good shot will often try out his position before he fires, to test this last point. He will close his eyes, bring his rifle up to the aiming position and then see where it is pointing. If it is not pointing at the target he will then come down, re-adjust his whole body and repeat the process until it is. The telescopic sight is a help in minimizing unsteadiness, because it reveals little movements

that could not be seen with the aperture sight. Make a firm base for the rifle with the body and elbows, and do not have the elbows too close together or this will make for lateral instability. Give the rifle firm support with both hands, shoulder and chin, and most strongly with the right hand, which should be closed completely round the small of the butt, pulling inwards and back against the shoulder and chin. In no case should strain be involved, or this will inevitably cause unsteadiness. But a strong grip of the right hand is of great importance, and it must be stressed, more particularly since most men are inclined to hold loosely at this point. Another common fault is holding with the left hand close back to the magazine. The left hand should be as far forward as it will comfortably go, with the rifle resting right down in the palm. Good holding comes from practise off the firing point as much as on it. The sniper should practise holding onto a small mark, and try out different positions, to see how steadiness can best be achieved.

89. Flinching.—This is a reflex action in anticipation of the shock of discharge. It may be caused by gun-shyness, or more commonly by a bruised cheek or shoulder through loose holding. It is a difficult fault to detect, because it is often hidden by the rifle's discharge, but it will cause wild and scattered grouping. In cases of doubt it can often be detected by making a man fire with an unloaded rifle, but the coach must not allow the firer to know that his rifle is unloaded, otherwise he will be unlikely to flinch. Even an experienced shot may sometimes fall into the habit of flinching, but if so he should be able to detect it in himself. The cure can generally be effected by conscientious snapping, until the fault is eliminated. The inexperienced shot will, however, need the help of his coach, using an eye disc, and also perhaps a bit of persuasion, if nervousness is the cause. If flinching arises from a bad bruise, it is better, if possible, to leave shooting alone for a few days or the fault may only grow worse.

90. Faulty trigger let off.—The sniper must be able to let off his trigger instantaneously and without disturbing his hold, as soon as his eye tells him that his aim is correct. If he does disturb his hold, or, in other words, snatches the trigger, a bad shot will naturally result, usually going low and to the right. On the other hand, if his hold is really firm, and the right hand has a strong grasp of the butt, it is virtually impossible for him to snatch, particularly if he has a good control over his trigger finger. An instantaneous trigger let off is necessary, because it is not possible to hold a rifle perfectly still, and unless the shot is let off at the moment that the eye proclaims the aim to be correct, the rifle will almost certainly have moved off aim. The best shots are nearly always the quick ones, for the longer the time spent in the aim, the more will the eye and the muscles have tired and the smaller will be the chance of a good shot.

The trigger should be held low down towards its free end, in order to obtain the best leverage, using the middle of the finger, rather than the tip. In this way the finger will be the least sensitive to any drag or hardness of the second pressure. Drag can, however, be very putting-off ; and since it is easily remedied by an armourer, it should, if detected, be dealt with at once in the armourer's shop.

Test a man's trigger let off by making him take a dry snap with a sixpence balanced on the muzzle of his rifle, and see if it remains there.

91. Faults in aiming, such as varying the amount of foresight, or inclining the sights are, together with their results, so well known, that it is not necessary to discuss them here. They should never arise in a trained soldier with normal eyesight.

92. Incorrect breathing.—The fault here is likely to lie, not so often from failure to stop breathing before firing, as from holding it too long. This will cause strain and unsteadiness. The breath should be held for as short a time as possible before firing the shot, and with the lungs partly deflated.

93. These points should receive the constant attention of the coach during the initial firing practices. Much can also be done to improve them in short periods of dry snapping. When they have been mastered and preferably not until this is so, the sniper can begin to consider the outside influences that will affect his shooting. If his grounding in this is secure, he can begin to tackle the other problems with confidence.

Section 20 Sources of error in the rifle and ammunition

94. Construction of a service cartridge.—A Mark 7 cartridge is in itself an instrument of great precision, evolved from much experiment and good workmanship. It consists of a rimmed cartridge case of solid

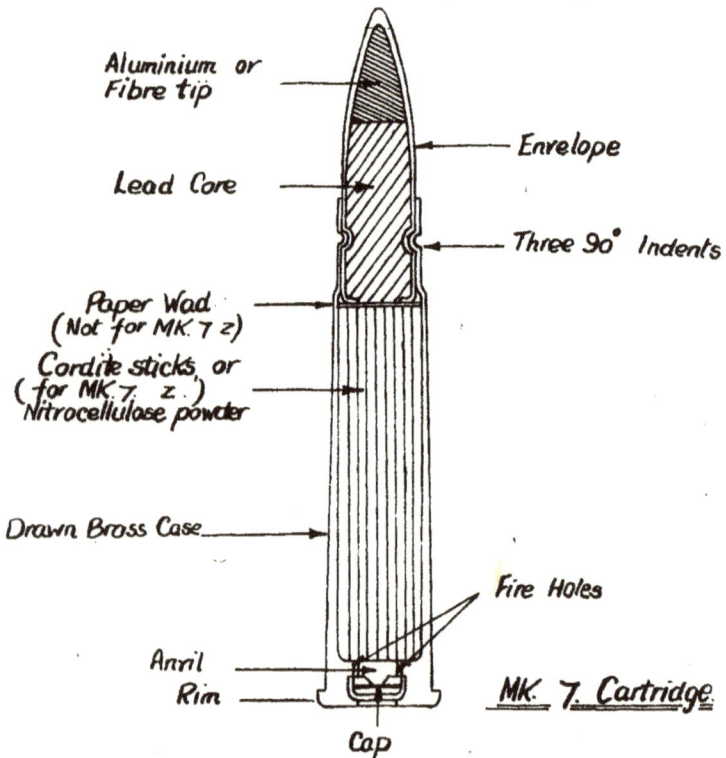

Aluminium or Fibre tip

Lead Core

Paper Wad (Not for MK. 7 z)

Cordite sticks, or (for MK. 7 z.) Nitrocellulose powder

Drawn Brass Case

Anvil

Rim

Cap

Envelope

Three 90° Indents

Fire Holes

MK. 7 Cartridge

drawn brass, varying in hardness from top to bottom. The case is filled almost completely with cordite sticks, with the bullet crimped tightly into the narrow neck of the cartridge. Into the cavity in the base of the cartridge is inserted a soft copper cap, which has been filled with a small quantity of cap or priming composition. The bullet itself consists of a tough cupro-nickel or gilding metal envelope, and inside the point of the envelope is pressed an aluminium or compressed paper tip. This light tip helps to balance the bullet, the remainder of which is filled with lead.

95. Action on firing.—When the trigger is released, the striker is forced forward by the spring inside the bolt and through the hole in the face of the bolt, punching a dent in the soft copper of the cap. The following things then happen :—

(a) The priming composition is detonated by the punch of the striker, and the flame from it shoots out through fire holes, connecting the cap to the inside of the cartridge. This flame fills the interior of the cartridge and sets fire to the cordite.

(b) The gases caused by the primer and the burning cordite begin to set up a high pressure inside the case. The higher this becomes the faster the cordite burns, all the time generating more and more gases at a faster and faster rate, until the pressure becomes such as to expand the soft brass at the front of the case and drive out the bullet.

(c) The gases have now a chance of escape behind the forward moving bullet, which is forced at an ever-increasing pace up the barrel. At the same time the bullet is squeezed into the spiral rifling of the barrel, and has to revolve in the direction of the rifling, until it leaves the muzzle at a speed of about 2,440 feet per second, spinning at 3,000 revolutions per second. This spin keeps the point of the bullet foremost during flight.

(d) The maximum pressure developed by the gases comes soon after the bullet enters the barrel, when it reaches $19\frac{1}{2}$ tons per square inch. At this pressure the slightest leak round the sides of the cartridge or the bullet itself would lead to some of the gas escaping without doing its job, but this is prevented by two things. The first is that the great pressure slightly squashes the base of the bullet, so that it expands and fits the rifling of the bore very tightly. This action is known as set-up. The second is that the cartridge expands and grips the walls of the chamber, also very tightly. This is important, not only because it makes a seal, but also because it relieves much of the pressure coming back on to the bolt. The whole of these operations between the time the trigger is released and the bullet leaves the muzzle have taken some time to describe, but in fact take $\frac{7}{1000}$ of a second.

(e) Two other things of note, one of them very important, take place during the explosion. First the rifle itself moves backwards, recoiling from the bullet's forward movement. Most of the recoil is only felt after the bullet leaves the muzzle. Until it reaches the muzzle, the rifle, if held properly, moves back about $\frac{1}{10}$ of an inch, so this movement has practically no effect on the result of the shot. Slight differences only may be noticeable if the butt is moved up or down on the shoulder between shots.

(*f*) The second is more important. The force of the explosion makes the barrel vibrate, so that the bullet leaves the end of the barrel during these vibrations, and, therefore, not in the direction in which the barrel was pointing as the trigger was pressed. This is called " jump." Fortunately, in any one rifle, the bullet always leaves at almost exactly the same point in the upward or downward vibration, and jump is, therefore, compensated for when the rifle is zeroed. But the real point of importance is that, if certain quite trivial things are done to the rifle, the normal vibrations of the barrel will be altered, the jump will be different and a bad shot will result. This point will, therefore, be considered later.

96. Trajectory.—After the bullet has left the barrel, the air resistance begins to slow it down immediately and at the same time gravity begins to pull it down towards the ground. These two forces make the path of the bullet curve downwards, very slightly at first, but more steeply the farther it goes. The slowing down of the bullet is very great, so that although it leaves the muzzle at just over 800 yards per second, in the first second of its flight it only in fact goes 600 yards, and in the second second only 400 yards.

The shape of the curved path of the bullet, or its trajectory, described above, explains why the elevation required on the sights at the longer ranges is so much greater than that needed at the shorter ones. But notice that if a bullet were fired straight up in the air, or straight down from a balloon, the path would not be curved, so that no elevation would be needed at any range. The practical aspect of this fact is that, when fighting in hilly country, and firing up or downhill at a steep angle, less elevation is needed than usual. The steeper the angle, the less is the elevation required, at 40 degrees the elevation needed being only three-quarters that of normal.

97. Sources of error in the rifle.—Most of these are connected with some interference with the normal jump. It is only possible to list the more important and most usual of them.

(*a*) A neglected barrel will cause increased friction and, therefore, decreased accuracy. A sniper should take a pride in maintaining his barrel in a spotlessly clean condition. A simple way of doing this, even in the field, is given in Sec. 22. Once neglected, the original polish of the bore can never be fully regained.

(*b*) An oily barrel has much the same effect as a neglected one, at any rate for the first two or three shots, which are the ones that count in war.

(*c*) Oil in the chamber has a worse effect than in the barrel, since it will continue to give bad results for many rounds. Its effect is to prevent the cartridge from gripping the sides of the chamber, which alters jump, and also causes abnormal pressure on the bolt and action of the rifle.

(*d*) Oil on the face of the bolt may also alter the normal jump.

(*e*) Nickelling, or patches of nickel scraping off the bullet envelope on to the lands of the rifling, will cause inaccuracy. The effect of this will be very slight, but a watch should be kept for its appearance, particularly in a worn barrel. Nickelling can be seen as rough, light-coloured streaks generally on the lands. If it becomes bad it should be removed by an armourer.

(*f*) A bulged barrel will impair shooting to varying extents. Bulges can be caused by very small amounts of dirt, sand or snow entering the barrel, before firing. Very great care must, therefore, be taken to prevent this happening when the sniper is crawling.

(*g*) The barrel of the No. 4 rifle is a floating barrel. That means that the woodwork does not touch it anywhere, except for a short distance at the muzzle and chamber ends. If the woodwork becomes warped and bears on the barrel, this will affect jump. Wet or heat may cause warped woodwork, unless this is regularly oiled (*see* A.C.I. 1148/40). In the No. 4 rifle the barrel should bear downwards on the woodwork at the muzzle end, with a pressure of 3–4 lb., and it should have a slight play in an upward direction.

(*h*) A bayonet should never be fixed to a sniper rifle, since this will greatly affect accuracy.

(*i*) Loose screws and particularly the one in front of the magazine will alter jump, and should be checked periodically.

(*j*) Resting the rifle on something hard will cause an alteration in the position of the M.P.I., depending on the point at which the rifle is rested.
Since the sniper may have to do this in an emergency, he should test its effects on his elevation.

(*k*) Some barrels will fire low when they become heated by firing. This also should be tested out, though a hot barrel should seldom arise in the sniper's case.

(*l*) A loose butt will cause bad shooting.

(*m*) A bolt not belonging to the rifle may also do so.

(*n*) The foresight or aperture backsight may become loose or bent, and should be regularly examined. Difficulty in aiming may also be found if the foresight is in want of browning.

98. Sources of error in the ammunition

(*a*) Oily or dirty ammunition will have the same effect as oil in the chamber.

(*b*) Wet ammunition will cause high shooting. During rain, ammunition must either be kept dry or allowed to become thoroughly wet and 100 yards deducted from the elevation.

(*c*) Dirt or sand on the bullet will cause a bad shot, and will also scratch or bulge the barrel.

(*d*) Ammunition that becomes hot, either from lying in the sun or from being left in an overheated chamber, will shoot badly.

(*e*) Different batches of ammunition, even if made at the same factory may have slightly different muzzle velocities, and, therefore, vary in elevation. A sniper should stick to one batch for as long as he can, and check his elevation if he has to change it.

(*f*) Armour-piercing and tracer ammunition have different trajectories from that of ball. Tracer and armour-piercing are also less accurate than ball, and wear out the barrel very quickly; tracer is very corrosive. The immediate firing of two or three rounds of ball will remove this corrosive residue.

Section 21   Zeroing, application of group and coaching

99. Zeroing and error of the day.—The greater the skill of the firer, the more important becomes the accurate zeroing of his rifle, because a moderate shot with a wide grouping capacity could hope to hit the target with some of his wide shots, even though his rifle is somewhat out of zero, but a consistently close grouper would put all his shots off the mark in such a case. For this reason, the initial zeroing of both aperture and telescopic sights of a sniper's rifle, must leave only the smallest possible error on the sights. The mechanics of this initial zeroing are described in Sec. 115, but this is not all. Owing to an accumulation of minute differences in replacing the telescopic sight on the rifle and differences of hold, and/or atmospheric conditions, the sniper must realize that no rifle can be relied on to maintain its zero perfectly from one day to another.

Effect of badly zeroed rifle for good and bad shots

Point of Aim

M.P.I. of Rifle incorrectly Zeroed

Group made by good shot

Group made by poor shot (Some shots may hit shaded area of man.)

(Target at 500 yds)

100. Application.—The object of application practices is to give the sniper a chance of learning how to overcome these differences, as well as the factors of wind and light, and for this reason he must receive careful coaching in the use of sighting shots. Two sighters should be allowed before every application practice, so as to determine the error of the day. If these sighters do not fall within the grouping capacity of the firer, the best course to take, provided that both shots were declared correct, is to calculate their M.P.I. or the point midway between them, and alter the sights or the aim so as to bring the shots central from this point. It is seldom advisable to make any correction on the first sighter, since, excepting where this is very wide, it gives little indication as to the ultimate position of the M.P.I. By dividing the distance between the two, however, a reasonably accurate indication of this will be gained. The exception to this rule will come where the first sighter, declared correct, falls near the edge of the target. In this case it is best to make a moderate alteration, otherwise there is a chance of the second sighter being off the target, when its value will be lost.

101. Examples.—The following examples will help to illustrate these principles :—

(a) A 2½-inch grouper at 100 yards is firing at 200 yards at a target with a 6-inch bull and 12-inch inner. His sighters, A and B, fall as follows :—

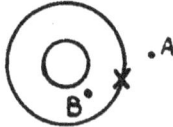

X, the M.P.I. of these two shots, is 6 inches right of centre and 6 inches correction to the left is, therefore, needed.

(b) A 2½-inch grouper at 100 yards is firing at 500 yards, on a 4-foot target. The first sighter falls in the outer at 9 o'clock, and is declared correct. The sights, or aim, should be brought over approximately a foot towards the right, before firing the second sighter, and this must be allowed for in calculating the M.P.I.

(c) A 3-inch grouper is firing at 200 yards at a target with a 6-inch bull, and a 12-inch inner.
The sighters, A and B, fall as follows :—

X, the M.P.I., is approximately 2 inches low, and is inside the firer's grouping capacity. No alteration is needed.

(d) A 2½-inch grouper at 100 yards is again firing at 200 yards. The sighters fall as follows, *B being declared low right.*

Here, the position of X is an unreliable guide and a cautious alteration, acting on the position of A, is the best policy.

102. Coaching.—It will be noted that the principles as to alteration of sights or aim vary in no way from those given in Small Arms Training, Vol. I. Any apparent variance between the coaching of snipers and other soldiers is because the grouping capacity of the former must be so much smaller than that of the ordinary soldier, that alterations may be needed, even where the error is comparatively small. This fact may easily be overlooked by the instructor, who is used to handling non-snipers on the range, and who may, therefore, not demand a sufficiently high standard of accuracy.

103. The principles given must, in any case, be applied with common sense, guided by experience. For instance, in the example (c) of para. 101, the two sighters, although within the grouping capacity of the firer, might well be the central shots of the group and not the low ones, as has been assumed. If this were so, an alteration of 2 inches up would be needed, to prevent some shots dropping into the inner. Other cases will occur where a man may think both sighters have been well let off, and yet they are wider apart than his normal grouping capacity. In this case the coach may have been able to detect a faulty let off by the firer in one of the shots, which he can, therefore, discount. Otherwise, experience alone can give the answers. The coach must be prepared for an unaccountably wide shot and to answer the questions—was it a bad let off ? Did the wind change before the shot was fired ? Did a sudden change in light affect elevation ? Were the sights incorrectly set ? or Has the rifle developed a fault ? and many other such pertinent questions. Sometimes the subsequent shots fired may seem to bear no relation to the position of the sighters. Here again the coach has a problem which he must work out, carefully searching for the reasons. If he alters, he may find himself chasing errors. On the other hand, the sighters may have been in some way faulty.

It will be seen that the task of coaching to the standards required calls for a great deal of practice and concentration, and the sniping instructor must demand of himself a high standard in this.

104. **Score cards.**—For practice shoots on the range, it is essential that a detailed record should be kept, if the full benefit is to be obtained from them. Army Forms B2141, which are special score cards, issued for snipers, should, therefore, be used on all application practices. On them is recorded the strike of each shot on the target, with the elevation and lateral allowances on the sights. These must be entered up, shot for shot, as it is signalled, either by the firer or his coach. This record will help the coach, and will also be of value to the sniper for future reference in getting to know his rifle.

105. **Snapshooting.**—Although they give less chance of studying the application of group, a large proportion of snapshooting and moving target practices, must be included in the sniper's training. Through these he will learn to become a quick shot, at the type of targets met with on service.

106. **Maintenance of zero in the field.**—It is of even greater importance in the field, where his life may depend on it, that the sniper should keep his rifle in zero. This can often be done in battalion rear areas, using some improvised mark. Occasionally, choosing a moment when other noises will cover that of his rifle's discharge, the sniper may fire one or more ranging shots from his actual battle position. This he will do in a puddle of water or a patch of sand, or at anything that will show the bullet's strike. But care must be taken not to shoot at anything near the enemy positions which will give the sniping post away. The direction of bullet holes through a tin can will be noticed by a wary enemy. He may even have placed the can there on purpose, a ruse which can sometimes be used to trap enemy snipers. Indiscriminate firing of shots is to be avoided in the forward area, but some means must be found of keeping this constant check on zero.

107. **Flattening the trajectory.**—Despite the essential fact that a sniper must be able to adjust his shots in inches, on service, he may have to take a very quick shot, without having time to adjust his sights.

Because of this the sniper should make a habit of carrying his rifle with the sights set for 300 yards. As can be seen from the accompanying diagram, this will ensure him a hit on a man's body at any range up to 400 yards, with little, if any, need to aim up or down. This stratagem is sometimes called flattening the trajectory.

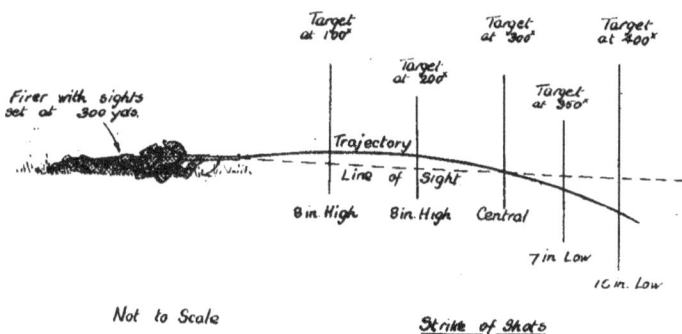

Flattening The Trajectory

Section 22   The No. 4 (T) sniper rifle

108. General

(a) The equipment consists of a No. 4 rifle fitted with two pads to take the telescopic sight No. 32, a cheek rest, a special sling, a cleaning cloth and a 32 telescopic sight in a special case, which may also contain tools adjusting No. 1, Mark 1 and No. 2, Mark 1. No tool is issued with the Mark 3 telescopic sights. The complete equipment is carried in a chest S.A. No. 15.

(b) In order to get maximum efficiency from the sniper rifle, particular attention must be paid to the fitting of the rifle to suit the individual. Whilst the positioning and height of the cheek rest is comfortable to a man of normal build, occasions arise when some slight modification is necessary to obtain maximum comfort. It will also be necessary to fit a long or a short butt, in the case of men having abnormally long or short arms. These special lengths of butt can be obtained through the normal channels. Such alterations to suit any particular sniper should be carried out under local arrangements.

109. Cleaning the sniper rifle.—Special care is needed if the sniper rifle is to retain the high performance of which it is capable. This applies particularly to the matter of cleaning the barrel after firing. Whenever possible, this should be done by " boiling out " in the normal way, but in the field, or at other times when boiling water is not available, the following method of cleaning is equally efficient, if carried out soon after firing.

(a) Pull the rifle through several times with a piece of flannelette soaked in water or saliva. In this way the corrosive fouling, which is insoluble in oil, will be removed. Saliva, or soapy water are even better than plain water for doing this.

(b) Dry out the barrel thoroughly with successive dry patches.

(c) Oil the bore with a patch measuring $4 \times 1\frac{1}{2}$ inches.

(d) Repeat the process on the following day, to prevent " sweating."

110. This method is simple and thoroughly effective, and can be safely employed by the sniper at any time. A still better clean will be obtained, if a cleaning rod is used instead of the pullthrough, and every sniping section should possess two or three of these. A gauze should never be used in a sniper's rifle, except when it has suffered from unavoidable neglect on active service.

Section 23   **The No. 32 telescopic sight**

111. The following points connected with the equipment should be noted :—

(a) The focus of the telescopic sight cannot be altered.

(b) The polishing cloth is for use in cleaning the lenses, and will always be kept clean.

(c) Telescopic sights are paired with rifles and are NOT interchangeable. Whenever a rifle is sent in for repair, the telescopic sight will always accompany it and *vice versa*.

(d) When fitting the telescopic sight to the rifle, first ensure that the number of the sight is the same as that on the rifle, and that the bracket and pads are free from dirt. Then partially tighten the clamping screws, giving the final tightening to the front screw first and the rear screw last.

(e) The magnification of the sight is 3.

112. **Adjustment of telescopic sight.**—Elevation is graduated from 0 to 1,000 yards and is adjusted in 50-yard clicks (Mark 1 sight), and 1 minute clicks (Mark 2 and Mark 3). The range is set when the figure for range is opposite the pointer mark. When going UP the scale, always move drum one click above the range required, and then come back one click. This eliminates any possible " back-lash " in the sight mechanism. When going DOWN the scale, the drum is set to the exact range required.

Lateral adjustment is obtained by turning the deflection drum on the left of the telescopic sight. It is graduated in 2-minute clicks (Mark 1 sight) and 1-minute clicks (Mark 2 and Mark 3), up to 16 minutes on each side of zero. A minute of angle equals approximately 1 inch on the target for every 100 yards of range, *e.g.* :—

Mark 1   Telescopic sight—2 clicks=4 minutes.

4 minutes at 300 yards=12 inches.

Mark 2 or 3   Telescopic sight—2 clicks=2 minutes.

2 minutes at 300 yards=6 inches.

If a shot strikes *left* of the mark, turn the drum anti-clockwise and *vice versa*.

113. Method of aiming

(a) Keep the sights upright. The crosswire is of assistance in this.

(b) Close the disengaged eye.

Correct Aim  **Floating Aperture**

(c) A full field of view must be obtained. If the eye is too near or too far from the sight, a floating aperture will be obtained.

(d) Place the tip of the pointer in the centre of the target.

114. Zeroing.—This consists in adjusting the range and deflection scale readings, so that the M.P.I. coincides with the point of aim, with the drums set at 1 and 0 respectively.

General considerations for zeroing are as follows :—

(a) *Examination before test.*—The rifle and the telescopic sight must be examined by the armourer before test, to ensure that all screws are tight and that the rifle is stocked up correctly and free from nickelling and oil.

(b) *Weather conditions.*—Good shooting light and the calmest available weather conditions must be chosen. Sights will be set at 100 yards.

(c) *Range.*—100 yards from foresight to target.

(d) *Position.*—Lying. Forearm and wrist rested (not the rifle).

(e) *Targets.*—Any suitable target with a *1*-inch black (or white) aiming mark.

(f) *Point of aim.*—Lowest central point of aiming mark.

(g) *Method.*—Two shots will be fired into the bank to warm the barrel. A group of five rounds will be fired. The necessary standard of grouping is 3 inches at 100 yards.

(h) After any adjustment has been made, the rifle should be re-tested and corrected as necessary.

Note. — The least possible vertical or lateral error should be left on a sniper's rifle.

115. Adjustment of errors

(a) First alter the sighting of the range and deflection drums, so as to bring the M.P.I. to coincide with the point of aim.

(b) *Marks 1 and 2 sights*

(i) *Using the new tools* (Tools No. 32 Telescope No. 1, Mark 1 and No. 2, Mark 1).
Ensuring that the central pin of the range drum is not disturbed :—

A. Apply the jaws of the No. 1 tool round the range drum and engage the projection on the central pin, with the slot of the plug.

B. Tighten the jaws and then, with the aid of the No. 2 tool, loosen the clamping ring.

(d) If it is impossible to put the elevation scale low enough to
make the M.P.I. coincide with the point of aim, the following
example will explain what to do:-

When the sight is set at 100 yards, the M.P.I. is found to
be approximately 8 inches high. It is thus necessary to lower
the sights 200 yards to bring the M.P.I. central. This is,
however, impossible. Therefore:-

(i) Using the Mk. 1 or Mk. sight. - Keeping the centre pin
still, loosen the clamping ring, adjust the range drum to
read 300 yards, and clamp up again. Then turn the drum
down to 100 yards.

(ii) Using the Mk.3 sight. - Holding the milled edge of the
elevation drum still, move the scale until it reads 300
yards. Then turn the drum down to 100 yards.

(iii) In both the above cases the sights have been moved down the
equivalent of 8 inches on the target, and they now read the
correct range required.

(iv) Fire a check group to ensure that the telescopic sight is
now correctly zeroed.

C. Ease the tension on the jaws slightly and make the necessary adjustment with the fingers, until the drum is set at the correct reading for the range.

D. Tighten the jaws, and clamp up using the No. 2 tool.

(ii) *Using the original tool* (an assistant to hold the range drum will make adjustment easier).

A. Place the tool over the range drum, so that the slot on the end of the upper tommy bar engages with the projection of the central pin, and the projections on the base of the lower tommy bar fit into the slots of the clamp ring.

B. Turn the lower tommy bar anti-clockwise one half turn to loosen the clamp ring, at the same time holding both the upper tommy bar and the range drum stationary.

C. Keeping the upper tommy bar stationary, rotate the range drum until the range to the target is shown opposite the pointer mark.

D. Holding the upper tommy bar and the range drum stationary, tighten the clamp ring by means of the lower tommy bar.

(iii) Corrections for direction are made by applying the above procedure to the deflection drum until the figure 0 is opposite the pointer mark.

(c) *Mark 3 sight.*—Fire a series of shots at 100 yards and adjust range and deflection drums until the M.P.I. of the group formed is at the point of aim.

No tool is required to effect adjustment with the Mark 3 sight; simply apply the nose of a bullet to the recessed projection on the range scales as follows :—

To zero elevation.—Hold the milled edge of the elevation drum firmly with one hand, and with the other, using the nose of a bullet, move the scale until it reads 100 yards elevation.

To zero deflection.—Repeat as for elevation, but move the scale until it reads zero deflection.

Care must be taken not to allow the drums to move when shifting the range scales.

Fire a check group, and should this not be correct the first time, repeat the process detailed above, until telescopic sight is correctly zeroed.

ection 24   Dusk firing

16. The telescopic sight enables accurate shooting to be done when is too dark to aim with the aperture sights. There comes a time very evening, when, with the aperture sights, it becomes no longer ossible to shoot at targets still visible to the naked eye. With the lescopic sight an accurate aim can be taken at anything that the aked eye is able to discern.

ther periods at which this advantage is marked are during moonlight r at dawn ; and since the enemy will hope to carry out many essential sks at these times, it is an important one.

117. Firing practices both at dusk or dawn, and in moonlight must be included in a sniper's training. These will not only give him confidence in his ability to do so, but will progressively improve his accuracy in this form of shooting. It should be noted that the contrast between the target and its background, rather than the state of the light, is the factor which governs its visibility. A target on a skyline can be aimed at in almost complete darkness. Experiments can be made with the visibility of various targets against different conditions of range and darkness.

118. It will also be found that, as in night observation, the clearest view of the target may be obtained by looking slightly away from it through the sight. With a little practice it is possible to take aim in this way without looking directly through the centre of the sight, and when this has been mastered it will give the best results.

Section 25   Aids to holding

119. **The sling**

(a) The sling can be utilized as an aid to holding, by transferring the rear end of the sling from the butt swivel to the swivel in front of the magazine. Instructions for fitting this swivel to the rifle are given in E.M.E.R. C.507, page 5.
In fixing the sling to this swivel it is important to see that the rough side of the leather is nearest to the rifle.

(b) To adjust the sling on the arm, hold the rifle with the right hand by the magazine, with the muzzle clear of the ground.

(c) Pass the left arm under the rifle from left to right, with the sling hanging loose. Then pass the left hand round the front part of the sling, in a clockwise direction, and grasp the rifle about the point of balance.

(d) Finally draw the rifle back slightly, and, with the right hand, hitch the sling well up the upper arm. Then assume the aiming position.

(e) If the sling is then too tight or too loose, re-adjust the length of the rear part of it, until a firm support is obtained, without undue pressure on the left arm.

(f) This method of using the sling can be quickly assumed or discarded in the lying position.

(g) When the sling is to be used purely for purposes of carrying the rifle, it is a matter of a few seconds to transfer its attachment back to the butt swivel.

120. **Hawkins position.**—When firing from behind low cover, great steadiness and good concealment can be obtained by using the Hawkins position. This consists in resting the left hand and forearm entirely on the ground, with the left hand holding the rifle about the outer band. The toe of the butt is placed firmly on the ground, its heel being under the firer's shoulder. When the cover is very low, extra elevation can be obtained by holding the sling where it joins the outer band with the clenched hand, as shown in the accompanying diagram. This hold can be so steady that it is barred from competition shooting. When used with the telescopic sight, great care must be taken to keep

Method of adjusting the sling.

Adjustment on arm.

Final position.

Hawkins position.

the eye back from the sight, or a cut eyebrow will result. It will also alter the normal elevation required on the sights, ~~50~~ to ~~100~~ yards less elevation being needed than usual.

121. Resting the rifle.—On service, the rifle will be rested, whenever possible, in the normal manner of firing over cover, with the wrist and forearm supported. This will, in fact, be possible quite frequently, while at other times added steadiness may be obtained by firing round the side of a tree, or even by holding a stick growing out of the ground. The sniper must be able to fire from any position that circumstances dictate, and also to take advantage of any aid to steadiness. For this reason, alternative firing positions must be included in his training.

Section 26   Elevation table

122. Introduction.—Small Arms Training, Vol. I, teaches that if shots are going low the sights will be put up 100 yards, and *vice versa*. For coaches, a somewhat more complicated table is given. This again is too rough for the sniper, who has to shoot to inches. This section gives an accurate table for use by snipers and must be thoroughly mastered.

123. The rule.—The rule governing the number of inches rise or fall, when raising or lowering the sights depends on :—

(a) The amount by which they are raised or lowered.

(b) The range at which firing is taking place.

124. For the Mark 1 sight the rules are as follows :—

(a) When firing at 100 yards the number of inches rise or fall on the target, produced by moving the sights up or down, is shown in the diagram below.

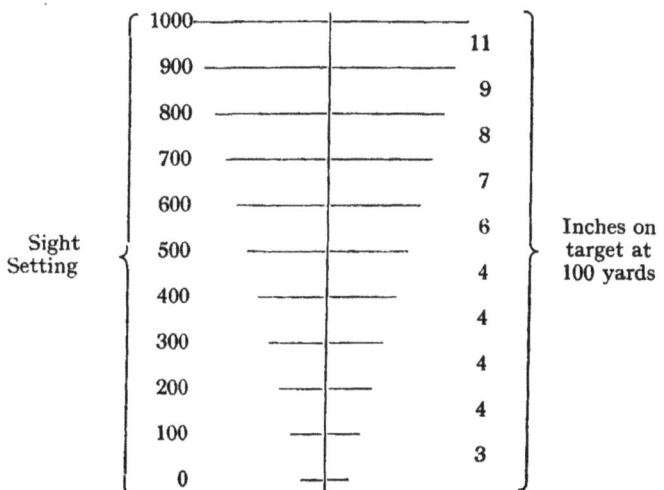

Sight Setting		Inches on target at 100 yards
1000		
900		11
800		9
700		8
600		7
500		6
400		4
300		4
200		4
100		4
0		3

(b) When firing at 100 yards and the sights are moved 50 yards only, the number of inches on the target will be half that for the whole 100 yards. Thus between :—

100 and 150 there are ½ of 4=2 inches.
150 and 200 there are ½ of 4=2 inches.
400 and 450 there are ½ of 4=2 inches.

(c) When firing at 100 yards and the sights are moved for more than 100 yards, the number of inches on the target is found by simple addition. Thus between :—

100 and 300 there are 4+4   = 8 inches.
100 and 350 there are 4+4+2=10 inches.
200 and 500 there are 4+4+4=12 inches.

(d) When firing at other ranges, the number of inches rise or fall in moving the sights is found, first by finding the amount this would give when firing at 100 yards, and then *multiplying* by the first figure of the range at which firing is taking place. Thus :—

When firing at 200 yards, if the sights are moved from 200–300, the result is $4 \times 2 = 8$ inches.
When firing at 300 yards, if the sights are moved from 300–450, the result is $(4+2) \times 3 = 18$ inches.
When firing at 300 yards, if the sights are moved from 300–200, the result is $4 \times 3 = 12$ inches.

125. The method of teaching for the No. 32 Mark 2 and Mark 3 sights can be simplified as follows :—

One click on the sights is equal in inches to the first figure of the range at which firing is taking place. Thus :—

When firing at 100 yards 1 click=1 inch.
When firing at 200 yards 1 click=2 inches.
When firing at 500 yards 1 click=5 inches.

126. The practical application of these rules for the adjustment of errors is as follows :—

(a) Determine the number of inches error on the target.

(b) Divide this by the first figure of the range at which firing.

(c) Apply the rule as given for 100 yards and move the sights accordingly.
It is pointed out that if the sights cannot be moved the exact amount required, it is, as a rule, better to move them to the next higher range, and aim down however much is necessary. It is easier to aim down than to aim up.

Section 27 Aiming off for wind and movement

127. Accurate wind allowance is one of the most important factors in good shooting at the longer ranges. Its effect on the bullet depends on the strength of the wind, the direction of the wind, and the range. Some means is, therefore, needed of estimating the first two of these.

128. Strength and direction of wind

(a) On the classification range there are often flags flying which will give a guide, both as to the strength and direction of the wind ; in the field, however, the sniper must learn to judge them from the feel of the wind on his face, or from the way in which smoke or foliage is being blown, or by the movement of ground haze or " mirage," as seen through his telescope.

53

(b) Bits of rag tied to bushes, remote from the sniping post, can also be a help, on occasions. By fluttering in the breeze they will not only help in judging the wind, but will also attract the enemy's attention away from the post.

(c) Wind strengths can be divided for practical purposes into three types—mild, fresh and strong. A mild wind will just keep a flag fluttering gently from its pole ; a fresh one will blow it out at an angle of 45 degrees, and a strong one will blow it straight out at right angles to the pole.

Mild   Fresh   Strong

(d) The direction of the wind is usually described by imagining the firer to be in the centre of a clock face, lying on the ground and with the target at 12 o'clock.

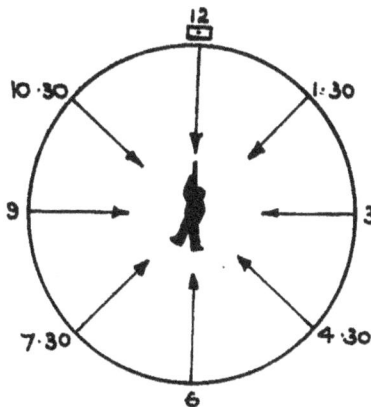

Notes.—(i) A 3 o'clock or 9 o'clock wind will have the greatest effect on the bullet.

(ii) A 12 or 6 o'clock wind will have no effect, except to vary the elevation by an amount that is negligible for practical purposes.

(iii) Other winds oblique to the line of fire will have an effect that becomes less and less, the nearer they get to 6 or 12 o'clock. Since it is not practical to list them all, commonsense is needed in estimating their effect, but when blowing from 1.30, 4.30, 7.30 and 10.30, they will have about half the effect of a 3 or 9 o'clock wind.

129. Wind table.—The following is a simple and easily memorized table that will give good results on service.

Effect of 3 or 9 o'clock wind

	Range	200 yards	300 yards	400 yards	500 yards	
Strength	Mild	3 ins.	6 ins.	9 ins.	1 ft.	¼ of fresh
	Fresh	6 ins.	1 ft.	1½ ft.	2 ft.	
	Strong	1 ft.	2 ft.	3 ft.	4 ft.	Double fresh

For oblique winds halve the above allowances.

130. **How to make allowances.**—Having calculated the wind allowance required, the sniper can either make the necessary adjustment to his deflection drum and aim central, or else aim off by the necessary amount. The latter course is strongly to be recommended, since in the field it is more practical to do this than to make sight adjustments. The telescopic sight, moreover, makes accuracy in aiming off comparatively simple. The sniper must be taught that the average width of a man's head is just under 8 inches, and of his shoulders it is 16 inches. These are the only guides he can expect on service in judging the allowance he must make. As in aiming off with the aperture sight, great care is necessary with the telescopic sight, to maintain correct elevation.

131. **Aiming off for movement.**—Up to a range of 300 yards it is quite possible to hit a walking, or even a running, man with good regularity. Practice and the aid of the telescopic sight are needed to do this. When the target is moving across the front it is necessary to aim off or lead the target, and to keep the rifle moving until after the trigger has been released. If the rifle is checked as the trigger is pressed, the benefits of the lead will be lost, and a miss behind the target obtained. Up to 300 yards the lead required for a walking man is half a figure's width, or his leading edge. For a running man it is necessary to aim off a full figure's width.

Section 28   Indication and recognition of targets

132. This section modifies Small Arms Training, Vol. I, Pamphlet 2, Chapter III, for indication and recognition of targets as between a sniper and his observer. A very high standard is demanded of snipers in this subject, because they must be able to bring accurate shots quickly to bear on individual targets. Team work and co-operation between the pair is essential for efficiency.

133. In the rifle section, the section commander gives fire control orders. For a pair of snipers the principles are the same as between a section commander and his section, with the following differences :—

    (a) The observer has only one man to whom to indicate the target.

    (b) The sniper's fire will be directed often at individuals in a group of men, rather than at the group, his object being to shoot key personnel, such as officers, N.C.Os., signallers, machine gunners, etc. A more detailed and accurate indication is, therefore, often needed, which must, at the same time, be accomplished quickly.

    (c) The target having been identified, the sniper is generally the best judge of when and how to fire, and he will not be dependent on the observer for any order to fire.

LESSON I.

1. Preliminaries.

(a) Stores:- 1 aiming rest per student. Students with rifles and
telescopic sights.

(b) Introduction. Explain that however good a pair of snipers may be
at stalking and shooting, they must also be expert at indicating
to each other the targets they find, and at picking up those
targets from the indication of them.

This lesson is slightly modified from the normal infantryman's lesson
since it only concerns a pair of snipers and not a whole section with
different weapons. One man will be acting as chief observer with the
Telescope Scout Regt., while the other, the sniper, will have binoculars.
They must both be able to indicate targets to each other.

Snipers have to pick out individual men as targets, and, if possible,
N.C.Os., machine gunners, etc. Accurate indication is therefore very
important.

2. Method.

(a) Give students a general line of direction, left of arc and right
of arc. Check by making them lay aims on these.

(b) Explain the sniper's fire order:- .

(i) Range. Stress importance of correct J.D.

(ii) Wind Allowance. So many minutes right or left.

(iii) Indication. Easy targets given by a line of direction.
Explain exact meaning of "General line of direction",
"slightly left", "quarter left", "half left", "$\frac{3}{4}$
left", "left", etc. Face squad on each direction as
follows:-

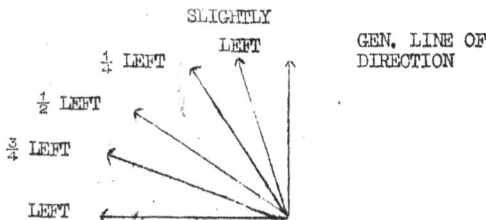

(iv) Confirmation. Sniper should say "O.K." if he has found
the target, but should wait, under normal circumstances,
until observer says "On" or "Fire now", before he fires,
so that shot can be observed.

(c) Give example of simple fire order:- "300 - wind drum 2 mins. left -
$\frac{1}{4}$ left - enemy right side of lone tree".
"O.K."
"Fire now".

(d) Practise squad in easy fire orders as above, squad instructor
giving fire order. Students to lay aims. Check aim and sights.

(e) Practise squad, in turn, giving easy fire orders, remainder laying
aims accordingly.

(f) Difficult targets. Further aids are required:-

(i) Reference points. See Lesson 7, para. 5(a). Question
squad on suitable reference points in each arc, and
judge distances to them. Explain that lines of dir-

3.

rection can be given from reference points where the latter exist, i.e. tree, slightly left, enemy, etc.

Practise squad in fire orders using reference points, squad instructor giving fire orders. Students to lay aims. Check aims and sights.

Practise squad, in turn, giving the fire orders, using reference points, remainder laying aims accordingly.

LESSON II.

1. Preliminaries.
   (a) Stores:- As for Lesson I, plus a pair of binoculars for each student.
   (b) Recapitulation. Practise squad in giving fire orders using reference points.
   (c) Introduction. Explain that further aids can be used to help indicate difficult targets.

2. Method.
   (a) Give students a general line of direction, left of arc and right of arc. Check by making them lay aims on these.
   (b) Difficult targets. Further aids can be used as follows:-
       (i) Clock ray method. Explain Lesson 7, para. 5(b) in detail. Practise squad in giving fire orders using this aid, remainder laying aims accordingly.
       (ii) Degree method. Explain Lesson 7, para. 5 c) in detail. Explain use of binoculars for this, using the division lines for accuracy.
          Practise squad in giving fire orders using this aid, remainder laying aims accordingly.
   (c) Emphasizes that in using any of these aids, or a combination of them, the indication of a difficult target should proceed stage by stage - from one prominent object to another; so many yards, etc.
   (d) Anticipatory fire orders. Sniper's example - "400" - wind zero - Farm house - right 3 o'clock - trench - enemy just visible at right end - keep on aim - await my order to fire".
      Practise squad in giving such fire orders, remainder laying aims accordingly.
   (e) Summarize main lesson taught.

LESSON III.

Preliminaries.
   (a) Stores:- 1 aiming rest per squad. Students with rifles, telescopic sights, Telescopes Scout Regt. and binoculars, 3 small and 3 large field firing targets. The large ones should be placed out in a position where they can be seen with the Telescope Scout Regt. or binoculars, but are difficult to distinguish with the No. 32 Telescopic Sight. The small ones should be placed in a position where they can be seen with the Telescope Scout Regt., but not at all with the No. 32 Telescopic Sight.
   (b) Recapitulation. Practise squad in giving fire orders, using aids so far taught in Lessons I and II.
   (c) Introduction. Very difficult targets. Explain that targets may often be seen with the Telescope Scout Regt., or with binoculars, but which cannot be picked up easily with the No. 32 Telescopic Sight.

In this case if the observer has seen a target which the
sniper cannot see, he should endeavour to indicate that target
by looking through his own No. 32 Telescopic Sight; both observer
and sniper will now be looking at the ground with the same
magnification, and this will enable the observer's indication
to be understood quicker and more accurately.

(d) Method. Squad should be split up in pairs in firing position.
Squad Instructor should call the observers up from each pair and
indicate one of the large targets to them by means of a Telescope
Scout Regt. on an aiming rest. Observers then return to their
snipers and indicate targets to them in the method explained
above. Squad Instructor criticises.

(e) Explain that some targets may be seen through the Telescope Scout
Regt. but which cannot be seen through the No. 32 Telescopic
Sight at all. This, however, does not necessarily mean that
target cannot be engaged. The same method as above should be
used, and the position of the target described accurately from
some prominent object close to it.

Example:- "500 - wind drum 3 mins. right - ruin - left 10
o'clock - 2 degrees - hedgerow - left 2 yards - gap - target one
foot right of gap under hedge".

"O.K."

"Fire now".

Practise squad in indicating the smaller targets by using
this method."

following sequence is all that the sniper and his observer

?ange.—This must be judged very accurately. An over-
stimate of 50 yards, when firing at 500 yards, will give an error
in the target of one and a quarter feet.

ndication.—As for fire unit, but must be more selective, *e.g.*,
**£50 — gateway — left 3 degrees — enemy section —
second man from left — a N.C.O.** or
' **200 — Bridge — Right 2 o'clock — Five small tufts of
rass — Right-hand tuft — Enemy sniper** ''.

\ind of fire.—Always single rounds. Usually the sniper himself can best decide when and how to fire. Sometimes, though,
the observer alone will be able to see when a target is exposed
at a given spot, and then he may give the executive word
'' **Fire** ''.

? in indication and recognition must frequently be given on the
and can be included in field firing practices.

le model landscape, constructed on a miniature range, with
aring targets, can also give very good practice in a convenient

,n 29 Field firing exercises

ld firing practices have the object of exercising the sniper
abination of observation, fieldcraft and shooting under field
ns. They form a culminating stage in his training. Ingenuity
needed in adapting them to the ground and facilities available,
here no field firing range exists, useful practices may be staged
classification range. Some suggested exercises are set out below,
ey will generally need some modification to suit local conditions.
amples given can do little more than provide ideas for the sniping
ctor. Some notes are also given on the preparation and conduct
exercises, for it is only by careful preparation and supervision
real benefit will be gained from them.

ractice I

f —Judging distance, indication of targets, observation, and
n ' fire between a pair of snipers.

ut.—A number of figure targets of varying sizes from 100 to 500
s from the fire position. These will be raised and lowered for
ing spaces of time, under control from the instructor at the fire
:ion.

hod.—Sniper will engage targets as they are exposed, the observer
:ating them to him if necessary, and giving him any correction
ed by observation of the bullet's strike.

munition.—10 rounds per pair.

Es.—Targets may be controlled :—

  (a) By having a fatigueman in a bullet-proof pit with each
      target. These are connected by lengths of string or
      cable with a N.C.O. or senior soldier in a central control
      pit, who signals to them by tugging on the appropriate
      string. The control pit will be in telephonic communica-
      tion with the firing point.

9. Page 55, para. 135.
Add at end "In all field firing practices, snipers should plot their own
position, and those of the enemy targets, on air photographs, from which,
in conjunction with a map, the exact distances between themselves and the
enemy targets can be ascertained. Due to the difficulty in judging distances, to the accuracy required for consistent results, by eye, this must
always be a most important consideration."

(b) By raising and lowering the targets themselves, with lengths of string, from one or more control pits connected by telephone with the firing point.

Time required : 45 minutes for two pairs.

137. Practice II

Lessons.—Stalking and shooting a selected target.

Layout.—A fatigueman, or student, is placed in a bullet-proof pit, close to which is the target to be stalked. He has with him a periscope and a pair of binoculars, also one Fig. 4a target and one Fig. 5, both mounted on long sticks.

Method.—Student " A " is shown the stationary target, near to the pit. He will stalk to within certain killing range of the target, and shoot it.

Student " B " is given a limited area to a flank in which to select a fire position covering the pit.

The fatigueman in the pit watches for student " A " through his periscope and if he sees " A " he will raise the Fig. 4a or Fig. 5 target, depending on how much he sees of him. He will lower it when " A " again disappears from view.

Student " B " will engage any target raised from the pit, and if he hits it, " A " is deemed to have been killed by " B." If " A " can stalk and shoot the stationary target, he is the winner.

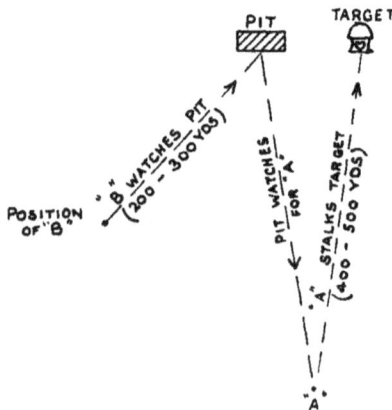

Ammunition.—" A " 5 rounds. " B " 10 rounds.

Time for complete exercise—1 hour.

138. Practice III

Lessons.—Taking up a concealed fire position and action in the position.

Layout.—An area is selected offering a number of fire positions that cover a bank or stop butt on which figure targets are placed. At the target bank and to a flank of the targets an instructor or selected student is situated under bullet-proof cover, from which he can observe with a periscope.

Method

(a) Students are paired off, and a number allotted to each pair. From a start-line in rear of the fire positions they will be shown the target area, and the area in which fire positions may be taken up. The flanks of this area will be clearly marked, and necessary steps taken to ensure that there is no possibility of students firing over one another's heads. This is most easily achieved if the forward edge of the cover forms a clearly cut line, parallel with the target bank.

(b) The object is for students to crawl forward to a fire position, and to search out and engage the target or targets bearing the number of their pair, and at the same time to avoid being seen by the instructor with the periscope. The numbers will be previously chalked on the targets, in figures just legible with a telescope from the fire positions. Points will be awarded at the end of the exercise for :—

(i) Hits obtained on the appropriate targets.

(ii) Movement into and actions in the fire positions, points being deducted from any pair who were seen through the periscope.

Ammunition.—5 rounds per pair.

Time required—45 minutes.

NOTES.—The time taken depends on the distance from start point to fire position. 45 minutes will be sufficient only if this distance is under 100 yards. By increasing it to a mile or more, the exercise can be made to embrace a lesson in cross-country map reading. Where this is done, the time will be increased accordingly.

139. Notes on the preparation of field firing exercises.—It is no exaggeration to state that at least two hours of preparation are needed for every hour of training. The following notes will assist officers or N.C.Os. in laying on field firing or fieldcraft exercises, and in obtaining the full benefit from the exercises themselves.

(a) *Preparation*

(i) The object of any exercise must be to bring out definite lessons which should be strictly limited in number. These lessons, once decided, must be constantly borne in mind, and the scheme developed around them.

(ii) Suitable ground must be found at an early stage. This may entail many unsuccessful attempts, before an entirely suitable area is discovered. Detailed reconnaissance is needed, to make sure that there are no unforeseen obstacles and to determine any boundaries required.

(iii) It is best, while still on the ground, to decide on the exact requirements of stores, or any work that will need to be done, and the number of fatiguemen required.

(iv) The exercise can now be written up in detail.

(v) Before putting on the exercise, any assistant instructors or fatiguemen to be used must be rehearsed in their duties.

(b) *Conduct of exercises.*—When writing out the scheme, the following points will, where applicable, need attention, so that they can be explained to the staff and students taking part.

(i) The lessons which it is intended to bring out.

(ii) The stores required.

(iii) The narrative, giving the tactical setting, if any.
(iv) Method of carrying out the exercise, in detail.
(v) Boundaries.
(vi) Time limits, if any.
(vii) Any signals to be used and action to be taken on them.
(viii) SAFETY PRECAUTIONS.
(ix) Administrative arrangements.

(c) *Discussion.*—The exercise should always end with a discussion, so that good and bad points can be brought to light. The discussion should take place immediately after the exercise. The longer it is put off, the less value will it achieve. The instructor will find it helpful during making his criticism, if he makes notes during the scheme on the execution of the following points :—

(i) *Observation*

Handling of the telescope and binoculars.
System in searching ground.
Quickness in location of targets.
Indication and recognition of targets.
Observation and correction of fire.

(ii) *Fieldcraft*

Reconnaissance before movement.
Route chosen, and why.
Method of progress.
Carriage of arms.
Use of camouflage.
Were risks taken, and if so were they taken early ?
Method of keeping direction.
Cover afforded by fire position or O.P.
Was good use made of cover ?
Did it give a good field of fire ?

(iii) *Shooting*

Weapon handling and sight setting.
Holding and trigger release.
Quickness in shooting.
Results on target.
Judging distance. ✓ use of air photographs ∨ maps
Allowance for wind and movement.

These points are not exhaustive, nor will all of them apply in every exercise.

APPENDIX A

SYLLABUS FOR BASIC TRAINING COURSE FOR SNIPERS

Subject	45 min. periods
Observation	
Lecture—General ...	1
Lesson—Telescope...	1
Lesson—Binoculars	1
Observation Scheme I	1
Observation Scheme II	1
Observation Scheme III	1
Long Range Observation Scheme	2
Demonstration—Location of Fire	1
Night Observation...	2

Subject.	45 min. periods

Employment of snipers

Lecture - Organization of sniper section	1
Film - "I'm a sniper" (shown twice)	3
Lecture - Employment of snipers	1
T.E.W.T. on sandtable.	4

Tests for Sniper's badge

1. Gen. knowledge	1
2. (a) Personal concealment	1
(b) Stalking (twice)	6
3. Observation	1
4. Judging distance (twice)	2
5. Map reading	3
6. Shooting (twice)	8

Miscellaneous

Opening talk	1
Gen. knowledge quiz	1 "

aft

General	1
How to Move	1
tration — Personal Concealment (without uflage)	1
stration — Personal Concealment (with uflage)	1
in concealment	1
tration—Stalking	1
nal Stalks	2
ction to Night Stalking	1
ual Stalks by Night	1
from Hides and Buildings (lecture or nstration)	1

Care and Cleaning of Rifle	1
Telescopic Sight I (description and aiming)	1
Elevation talk	1
Telescopic Sight II (Zeroing)	1
Aiming off for Wind and Movement	1
Shooting I (holding, aiming, trigger release, ation of group, sighters and score cards)	2
Judging distance (revision period)	1
Indication and Recognition of targets	1
Shooting II (sources of error in the rifle and aition)	1
Use of the sling, and Hawkins position	1

1. Range Practices. A suggested syllabus is given in Appendix B.

2. Field Firing Practices. These will depend on the time and resources available.

ading, Compass, and Aerial Photographs

ading Lesson preceded by Training Films—" X the spot " and " Know your Way "	1
based on " Hints on Map Reading Instruction ...	2
c Compass	1
Aerial Photo. Reading	1
Map Reading	2
March	1
Photo. Reading	1

aneous

talk	1
re—Enemy Badges of Rank	1
re—Tactical Handling of Snipers	1
re—Reporting and Panoramic Sketching	1

.—This syllabus is intended as a guide to the planning of an initial course of sniper training, lasting for less than two weeks. It allows for little revision or consolidation, and practice in the subjects covered can be profitably extended over any period of time available.

APPENDIX B

RANGE PRACTICES

The following practices are suitable for sniper training :—

Practice (a)	Detail (b)	Range (c)	No. of rds. (d)	Firing position (e)	Scoring (f)	Remarks (g)
1	Grouping	100 yds.	2 warmers 6 rds.	Over cover	2½-in. group 25 points 3 " " 20 " 4 " " 15 " 5 " " 10 " 6 " " 5 "	Target: 4 ft., with four 2-in. white aiming marks. Two 3-rd. groups will be fired at 2 separate aiming marks. Average size and position of both groups to be taken for scoring and zeroing purposes respectively. Correctly declared wide shots will be ignored for coaching and zeroing purposes.
2	Application	200 yds.	2 sighters 5 rds.	Over cover, or in the open, or with sling, or Hawkins position as ordered.	6-in. central 5 points 12-in. bull 4 " 24-in. inner 3 " 36-in. magpie 2 " 48-in. outer 1 " Possible 25 "	In the early stages of training, the sniper may be allowed to fire over cover, to help him to gain confidence. Later, suitable alternative positions should be used to an increasing extent.
3	Snapshooting	200 yds.	5 rds.	As for Practice 2	Bull 4 points Remainder 3 " Possible 20 "	Target: Fig. 4a, with 12-in. inscribed bull. 3 sec. exposures.
4	Snapshooting	200 yds.	5 rds.	As for Practice 2	Hit 3 points Possible 15 "	Target: Fig. 5, making the practice a more advanced one than Practice 3.
5	Moving target	200 yds.	5 rds.	As for Practice 2, except that Hawkins position is unsuitable.	Hit 3 points Possible 15 "	Fig. 4a target, moved at walking pace over a distance of 12 yds., from left to right, and right to left alternately.
6	Application	200 yds.	2 sighters 5 rds.	Firing over cover from left shoulder.	6-in. central 5 points 12-in. bull 4 "	Object of this practice is to give the sniper confidence in his ability to fire from the

RANGE PRACTICES

The following practices are suitable for Sniper Training:-

Practice (a)	Detail (b)	Range (c)	No. of rds. (d)	Firing position (e)	Scoring (f)	Remarks (g)
1	Grouping	100 yards	2 warners 10 rounds	Over cover, if zeroing. At other times using sling, either with or without stick-rest.	4 in. group 25 points 3 " " 20 " 4 " " 15 " 5 " " 10 " 6 " " 5 "	Target - 4 foot, with four 2 inch white aiming marks. Two 5 rd. groups will be fired at 2 separate aiming marks. Average size and position of both groups to be taken for scoring and zeroing purposes respectively. Correctly declared wide shots will be ignored for coaching and zeroing purposes.
2	Grouping	200 yards	2 warners 10 rounds	Sling or Hawkins	7 in. 25 " 8 " 20 " 9 " 15 " 10 " 10 " 11 " (one wide 5 points)	
3 X	Snapshooting	100 yards	5 rounds	As for Practice 5	Bull 4 points Reminder 3 " Possible 20 "	Fig. 5 target, with 4" inscribed bull, exposed for 3 secs., anywhere over a front of 10 yards.
4	Moving target	100 yards	5 rounds	As for Practice 5, except that Hawkins position is unsuitable.	Hit 3 " Possible 15 "	Target - Fig. 4a, moved at walking pace over a distance of 12 yards, from left to right, and right to left alternately.
5 X	Application	200 yards	2 sighters 5 rounds.	With sling and stick-rest, sling in the open, or Hawkins position as ordered.	5 in. central 5 " 2" bull 4 " 3" inner 3 " " magpie 2 " 4" outer 1 point Possible 25 points.	In the early stages of training, the sniper may be allowed to fire over cover, to help him to gain confidence. Later, suitable alternative positions should be used to an increasing extent.

8.

Practice (a)	Detail (b)	Range (c)	No. of rds. (d)	Firing position (e)	Scoring (f)	Remarks (g)
6	Snapshooting	200 yards	5 rounds	As for Practice 5.	Bull 4 points, Remainder 3", Possible 20"	Target – Fig. 4a, with 12 inch inscribed bull. 3 sec. exposures.
7	Snapshooting	200 yards	5 rounds	As for Practice 5.	Bull 4 points, Remainder 3", Possible 20"	Target – Fig. 5, with 8 inch inscribed bull, making the practice a more advanced one than Practice 6.
8	moving target	200 yards	5 rounds	As for Practice 4.	Hit 3 points, Possible 15"	As for Practice 4.
9	Application	200 yards	2 sighters 5 rounds	Firing over cover, from left shoulder.	6 in. central 5", bull 4", inner 3", magpie 2", outer 1", Possible 25"	Object of this practice is to give the sniper confidence in his ability to fire from the left shoulder, as he may have to do, on service, with certain types of cover.
10	Application	300 yards	2 sighters 5 rounds	As for Practice 5.	6 in. central 5", 12 bull 4", 24 inner 3", 36 magpie 2", 48 outer 1", Possible 25"	
11	Snapshooting	300 yards	5 rounds	As for Practice 5.	Bull 4", Remainder 3", Possible 20"	Target – Fig. 4a with 12" inscribed bull. 3 sec. exposure.
12	moving target	300 yards	5 rounds	As for Practice 4.	3", Possible 15"	As for Practice 4.
13	Application	500 yards	2 sighters 5 rounds	As for Practice 5, or using sling and cover.	12 in. bull 5", 24 inner 4", 36 magpie 3", 48 outer 2", Possible 25"	

Practice (a)	Detail (b)	Range (c)	No. of rds (d)	Firing position (e)	Scoring (f)	Remarks (g)
14	Application	600 yards	2 sighters 5 rounds	As for Practice 5, or using sling and cover.	As for Practice 13.	
15	Snapshooting	400 yards	2 sighters 5 rounds	As for Practice 5	Hit 3 points. Possible 15 "	Target - Fig. 2, 5 sec. exposure. No signalling of shots.
16	Application	200 yards	2 sighters 5 rounds each position.	Standing, squatting, sitting and kneeling, using sling; sawback and back positions.	6 in. central 5 " bull 4 " inner 3 " magpie 2 " outer 1 Possible 25	

NOTES:

1. These practices may be carried out with the aperture, or telescopic sights.

2. They are not arranged in what is necessarily the best order.

3. For application practices, the normal method of marking is not sufficiently accurate. After signalling the shot in the normal way, the markers should point to the exact shot hole, with the handle of the marking disc painted white, or should use spotting discs.

4. The 6 inch central, used for practices at 200 and 300 yards, is signalled, by placing the black side of the marking disc directly onto the centre of the target.

5. Sighting shots should be fired before snapshooting or moving target practices unless an application practice at the same range has taken place immediately beforehand.

6. Snipers should be worked in pairs for snapshooting and moving target practices, spotting each other's shots.

7. No conditions are set out for practices in dusk or moonlight firing, since differences of background will greatly alter the visibility of targets in different localities. Normally a Fig. 4a target at 100 yards is suitable for the first practice in dusk firing, but it is essential for the instructor previously to try out the target and range to be used, and to note carefully at what time the light will be suitable for carrying out the practice."

No.	Practice	Range	Rounds	Conditions	Scoring	Remarks
8	Application	300 yds.	2 sighters 5 rds.	As for Practice 2	Remainder 3 ", Possible 20 "	exposed for 3 secs., anywhere over a front of 10 yds.
9	Snapshooting	300 yds.	5 rds.	As for Practice 2	6-in. central 5 points, 12-in. bull 4 ", 24-in. inner 3 ", 36-in. magpie 2 ", 48-in. outer 1 ", Possible 25 "	Target: Fig. 4a with 12-in. inscribed bull. 3 sec. exposure.
10	Application	500 yds.	2 sighters 5 rds.	As for Practice 2	Bull 4 points, Remainder 3 ", Possible 20 "	
11	Application	600 yds.	2 sighters 5 rds.	As for Practice 2	12-in. bull 5 points, 24-in. inner 4 ", 36-in. magpie 3 ", 48-in. outer 2 ", Possible 25 ". As for Practice 10	
12	Snapshooting	400 yds.	2 sighters 5 rds.	As for Practice 2	Hit 3 points, Possible 15 "	Target: Fig. 2. 5 sec. exposure. No signalling of shots.

NOTES.—1. These practices may be carried out with the aperture or telescopic sights.

2. They are not arranged in what is necessarily the best order.

3. For application practices the normal method of marking is not sufficiently accurate. After signalling the shot in the normal way, the markers should point to the exact shot hole, with the handle of the marking disc painted white.

4. The 6-in. central, used for practices at 200 and 300 yards is signalled by placing the black side of the marking disc directly onto the centre of the target.

5. No conditions are set out for practices in dusk, or moonlight firing, since differences of background will greatly alter the visibility of targets in different localities. Normally a Fig. 4a target at 100 yds. is suitable for the first practice in dusk firing, but it is essential for the instructor previously to try out the target and range to be used, and to note carefully at what time the light will be suitable for carrying out the practice.

TEST FOR SNIPERS' BADGES

ırs' badges will be awarded to soldiers who pass the following tests :—

A.

ı) *Test 1. General knowledge.*—Answer two questions on each of the following subjects :—

    (i) The rifle. Mechanism, care and cleaning.

    (ii) The No. 32 telescopic sight.

    (iii) Elevation table.

    (iv) Aiming off for wind and movement.

    (v) Stalking telescope.

    (vi) Enemy identifications.

    (vii) Hides.

    (viii) Movement or observation by night.

Standard : 12 of the questions, and one on each subject, to be answered correctly.

ɔ) *Test 2. Stalking and personal concealment.* A practical test suited to local conditions, laid down by the battalion com mander who will be the authority for passing a candidate in this test.

ɔ) *Test 3. Observation.*—A practical test of observation, with stalking telescope.

Twelve articles of equipment to be partially concealed 100–300 yards from the observer's position. All objects to be within an arc of 30 degrees. Time for observation 30 minutes.

Standard : 8 out of the 12 objects to be located.

ɗ) *Test 4. Judging distance.* Practical test, judging eight ranges up to 1,000 yards.

Standard : Permissible margin of error—15 per cent. of the range.

ɛ) *Test 5. Map reading.*—Practical test. Sniper to find his way across 2 miles of country by map and without assistance.

B.

f) *Test 6. Shooting.*—See page 63.

13. Page 62, Appendix "C".

    (a) Para. (b)(i), lines 2 and 3, for "two or three minutes" read "up to five minutes".

        Para. (b)(ii), last sub-para. Delete last sentence and substitute "To pass the test the sniper must stalk to his final position and fire a round of blank ammunition at the enemy post before being seen by these observers."

    (b) Para. (c), line 4, after "position" insert "They must be placed out so that they are invisible to the naked eye, visible but indistinguishable with binoculars, but visible and distinguishable with the stalking telescope. "

        Line 6, after "to be located" add "and distinguished".

    (c) Test 5. Delete and substitute: "The sniper will be taken out to a spot in the open country which he has not seen before. He will then be told roughly his location by means of a 4-figure map reference.

        He will then be pointed out on the ground, six prominent features or objects between 300 and 1500 yards away, such as a house, a gully, a corner of a wood, a clump of trees, a corner of a field, etc.

        To pass the test he must, without assistance –

            (i) Plot his exact position on his map, and give a 6-figure map reference for it.

            (ii) Give the correct 6-figure map references of 5 out of the 6 landmarks pointed out to him, and the correct distance within 50 yards to each of these 5 landmarks from his own position.

        N.B. Compasses and protractors may be used."

14. Page 63, Shooting Test.

    (a) Top line. Delete "(Not included in classification)" and insert "(Qualifying practices.)"

    (b) Delete "PART II (Qualifying practices)" and insert "PART II (Classification practices)".

    (c) Practice 3.
            (i) Column 3 – For "200" read "300".
            (ii) " 6 – For "6-in. central" read "8-in. central".

    (d) Practice 5.
            (i) Column 3 – For "300" read "200".
            (ii) Remarks column – Delete and insert "The target to be moved over a distance of 12 yards in not less than 7 seconds and not more than 10 seconds. Any sniper firing when the target is not moving laterally will be disqualified."

    (e) Notes – For "QUALIFYING STANDARD" read "CLASSIFICATION STANDARD". Last line, after "initiative" add ", with the exception of sighters in Practice 5, which should be spotted for the sniper by someone else".

SHOOTING TEST FOR SNIPERS' BADGE—PART I. (Not included in classification)

Practice (a)	Detail (b)	Range (c)	No. of rds. (d)	Firing position (e)	Scoring (f)	Remarks (g)
1	Grouping	100 yards	2 warmers 3 rds.	Over cover		Average of 3-in. group or better before going on to Part II.
2	Grouping	100 yards	3 rds,	Sling or Hawkins		

PART II (Qualifying practices)

Practice (a)	Detail (b)	Range (c)	No. of rds. (d)	Firing position (e)	Scoring (f)	Remarks (g)
3	Application	200 yards	2 sighters 5 rds.		6-in. central 5 points 12-in. bull 4 ,, 24-in. inner 3 ,, 36-in. magpie 2 ,, 48-in. outer 1 ,, Possible 25 ,,	Target, Fig. 5, with 4-in. inscribed bull. Rifle in the aim. Target to be exposed for three seconds anywhere over a front of 10 yards.
4	Snapshooting	100 yards	5 rds.	Sling to be used	Bull 4 points Remainder 3 ,, Possible 20 ,,	
5	Moving target	300 yards	2 sighters 5 rds.	Firing over cover	Hit counts 3 points Possible 15 ,,	Target, Fig. 4a. Sighters to be fired at any mark on the bank. Target to be moved at walking pace over a distance of 12 yards.
6	Application	500 yards	2 sighters 5 rds.	Hawkins position	12-in. bull 5 points 24-in. inner 4 ,, 36-in. magpie 3 ,, 48-in. outer 2 ,, Possible 25 ,,	Target, 4-ft.

NOTES.—QUALIFYING STANDARD.—66 points out of 85. All practices to be fired with telescopic sight.
No assistance or coaching may be given in Part I or Part II and all corrections to sights or variation of aim must be made on the sniper's own initiative.

No. 32 TELESCOPIC SIGHT—CHART INSTRUCTIONS

The chart which should be obtained by unit armourer from R.A.O.C. under G1098 scale 4261, is intended as a means whereby a 32 sight, suspected of being faulty, can be checked up easily and the fault, if any, determined.

The following is the method of operation :—

1. Preparation

(a) *Rifle.*—Fix the rifle, with telescope attached, firmly in a vice or aiming rest, with the muzzle directed towards the wall or board on which the chart is fixed. The vice or aiming rest must be rigid, and, therefore, must be mounted on a solid base. A wooden floor should be avoided if possible ; any movement of the layer or other persons will upset the aim, due to vibration of the floor boards. If the ordinary aiming rest is not found to be rigid enough for accurate use of the chart, a rest made of sandbags should be used.

NOTE.—The correct method of attaching the telescope to the rifle is to screw up the clamping screws simultaneously, and to give the final tightening on the rear screw.

(b) *Telescope.*—Stop down the object glass with a card or tin disc, which has a hole bored in the centre not more than ¼ and not less than ⅛ inch in diameter. Stopping down the object glass eliminates parallax, which would otherwise be very troublesome at this short range.

(c) *Chart.*—Fix the chart :—

(i) Exactly 28 feet from the front lens of the telescope.

(ii) In a good light.

(iii) Horizontal. This can be checked by using a plumb line in conjunction with the central line of the range scale.

(iv) With the centre approximately the same height as the telescope.

2. To check the range drum

(a) Set the drum at zero and align tip of pointer on top line of vertical scale (marked 0 yards). Turn range drum through scale of ranges, checking each 100 yards with its equivalent line on the chart. Then, without adjusting the position of the pointer, re-check by working backwards to zero.

(b) Rotate the drum through full range and note that the tip o. the pointer does not move from side to side of a straight vertical path.

(c) To check for backlash. (This sub-paragraph does not apply to Mark 3 telescope.) Set drum at zero and align tip of pointer on top line of vertical scale. Turn range drum slowly up to its limit and back again to zero. Do this in two distinct continuous movements. Check to see if pointer returns to the same position, *i.e.*, tip of pointer on top line of scale.

3. To check the deflection drum

(*a*) Set drum at zero and align tip of pointer on centre line ot deflection scale. Turn drum and check each 2 minutes of angle of left deflection drum, with equivalent line on chart. Then, without adjusting the position of the pointer, re-check by working backwards to zero. Repeat above for minutes of right deflection.

(*b*) Rotate drum rapidly through full scale of 32 minutes, and watch for any variation in height of pointer.

(*c*) To check for backlash. (This sub-paragraph does not apply to Mark 3 telescope.)

    (i) Set drum at zero and align tip of pointer on centre line of deflection scale. Turn drum slowly to its left deflection limit, and back again to zero. Do this in two distinct continuous movements. Check to see if pointer returns to the same position, *i.e.*, tip of pointer on centre line of scale.

    (ii) Repeat above for right deflection.

4. To detect faults in holder or pads.—With range and deflection drums at zero, align tip of pointer on centre of diagram (top right of chart).

(*a*) Remove telescope from rifle, and replace by holding the telescope at the front end. Screw up the clamping screws in the normal way, whilst still holding the telescope at the front, tightening the rear screw last. The pointer should still be aligned on the centre of the diagram.

(*b*) Remove telescope and replace, but this time hold the telescope at the rear end and tighten the front screw last. The pointer should still be aligned on the centre of the diagram. Any variation in alignment of the pointer after (*a*) or (*b*) have been carried out, denotes a badly fitting or worn holder, and/or pad(s).

Note.—These tests will not reveal faults in the telescope caused by the shock of recoil when firing.

5. Sentencing an equipment B.L.R.—An equipment should be sentenced B.L.R. if at :—

Para. 2 (*a*) there is a variation exceeding 50 yards at any range between pointer and equivalent line on the chart.

Para. 2 (*b*) the pointer moves more than one minute to the left or right of the vertical path. (This can be judged from the horizontal scale.)

Para. 2 (*c*) the pointer's final position varies more than 50 yards from its starting position.

Para. 3 (*a*) there is a variation exceeding one minute either way.

Para. 3 (*b*) the pointer moves more than one minute up or down from its horizontal path.

Para. 3 (*c*) the pointer's final position varies more than one minute from its starting position.

Para. 4 when the telescope is replaced on the rifle, the pointer is more than one minute from its former position. (Limit of one minute in each direction is marked in the diagram.)

W.O.
CODE No.

7011-2

26/G.S. Publications/1535

SMALL ARMS TRAINING

Vol. 1, Pamphlet No. 28, Sniping, 1946

AMENDMENTS (No. 2)

1. Page 48, para. 115, *add* new sub-para. (*d*) :—
Amdt. 2/April/1947

    (*d*) If it is impossible to put the elevation scale low enough to make the M.P.I. coincide with the point of aim, the following example will explain what to do :—

    When the sight is set at 100 yards, the M.P.I. is found to be approximately 8 inches high. It is thus necessary to lower the sights 200 yards to bring the M.P.I. central. This is, however, impossible. Therefore :—

        (i) Using the Mark 1 or Mark 2 sight—Keeping the centre pin still, loosen the clamping ring, adjust the range drum to read 300 yards, and clamp up again. Then turn the drum down to 100 yards.

        (ii) Using the Mark 3 sight—Holding the milled edge of the elevation drum still, move the scale until it reads 300 yards. Then turn the drum down to 100 yards.

        (iii) In both the above cases the sights have been moved down the equivalent of 8 inches on the target, and they now read the correct range required.

        (iv) Fire a check group to ensure that the telescopic sight is now correctly zeroed.

2. Pages 52 to 54. *Delete* Section 27 and *substitute* new section (pages 53 and 53A to 53c), attached hereto.

By Command of the Army Council,

L.c. B.B.Mud.

THE WAR OFFICE.
    11*th April*, 1947.

Amdt. 2/April/1947

Section 27 Allowances for wind and movement

127. Accurate wind allowance is one of the most important factors in good shooting, especially at the longer ranges. Its effect on the bullet depends on the strength of the wind, the direction of the wind, and the range. Some means is therefore needed of estimating the first two of these.

128. Strength and direction of wind

(a) Much practice is needed in judging the strength and direction of wind. In the earlier stages, to help snipers in this task, a number of flags should be placed at intervals down the range. The flags will also bring home to them the degree to which wind will change in direction and strength during a short space of time. When proficiency has been gained with the aid of flags it is essential that these should be removed, and equal proficiency achieved from the feel of wind on the face, its effect upon grass or foliage, and from the movement of ground haze or " mirage ", as seen through the telescope. On service, dust and smoke from shell and mortar fire will serve also as a useful guide.

(b) Pieces of rag tied to bushes, remote from the sniping post, can also be a help, on occasions. By fluttering in the breeze they will not only help in judging the wind, but will also attract the enemy's attention away from the post.

(c) Wind strengths can be divided for practical purposes into three types—Mild (5 m.p.h.), Fresh (10 m.p.h.), and Strong (20 m.p.h.). A mild wind will keep a flag fluttering gently from its pole ; a fresh one will blow it out at an angle between 45 and 90 degrees, and a strong one will blow it straight out at right angles to the pole, and even higher.

MILD   FRESH   STRONG

(*d*) The direction of the wind is usually described by imagining the firer to be in the centre of a clock face, lying on the ground, and with the target at 12 o'clock.

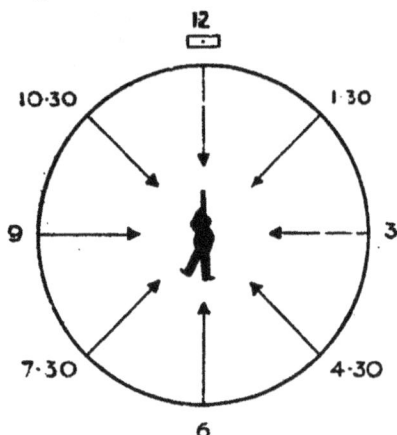

NOTES.—(i) A 3 o'clock or 9 o'clock wind will have the greatest effect on the bullet.

    (ii) A 12 or 6 o'clock wind will have no effect at normal shooting ranges.

    (ii) Other winds oblique to the line of fire will have an effect that becomes less and less, the nearer they get to 6 or 12 o'clock. It is not practical to list them all, but when blowing from 1.30, 4.30, 7.30 and 10.30 they will have about half the effect of a 3 or 9 o'clock wind.

129. **Wind table.**—The following table of allowances in Minutes of Angle is given as an accurate guide, and must be learnt by heart by the sniper.

Allowance for a 3 or 9 o'clock wind in Minutes of Angle

	Range in yards	100	200	300	400	500	600	
Strength	Mild (5 m.p.h.) ...	...	1	1½	2	2¼	3	(¼ of fresh)
	Fresh (10 m.p.h.) ...	1	2	3	4	5	6	
	Strong (20 m.p.h.)...	2	4	6	8	10	12	(Double fresh)

(*a*) For oblique winds halve the above allowances.

(*b*) To memorize the table the easiest way for the sniper to remember is that the number of minutes allowance for a fresh wind is the same as the initial figure of the range fired at in each case. This allowance is then halved or doubled for a mild and strong wind, respectively.

(c) ½-minute alterations are not possible on the deflection drum of the No. 32 Telescopic Sight. Therefore, where ½-minutes occur in the table, the sniper, if making a wind allowance on his sight, will have to set his sight to the nearest whole minute click possible.

(d) The above table has been proved by experience to be far more accurate than a previous table taught in inches and feet, which, though a rough guide and easy to learn, was not accurate enough for the sniper at the shorter ranges.

(e) It must be remembered that a wind of 20 m.p.h. or more is a comparatively rare occurrence.

130. Method of use

(a) *General remarks.*—There are two methods of making allowances for wind :—

   (i) By aiming-off a number of inches or feet into the wind.

   (ii) By making all allowances in minutes of angle on the deflection drum, and always aiming central.

Of these two methods the latter is by far the simpler, and will produce the more accurate results. Aiming-off a number of inches or feet is exceedingly difficult, and requires much practice ; this method does still have to be utilized, however, in certain circumstances with the Mark 1 sight, as explained below.

(b) *The Mark 1 sight*

   (i) With only 2-minute clicks on the deflection drum, small alterations are not possible on this sight. For instance, a reading of 1 or 3 minutes may be required as often as 2 minutes, and it is impossible to set the sight to the former two readings. It is therefore advisable when small allowances of up to approximately 9-12 inches on the target have to be made, for the sniper to convert the allowance in minutes (given in the table) to inches, and to aim that amount in inches on the target into the wind.
   An example :—300 yards, fresh wind blowing from right to left=3 minutes=3 × 3=9 inches. Therefore, aim right 9 inches.

   (ii) When a greater allowance is needed, and when, by keeping the deflection drum central, an aim-off of a number of feet would be required, it is far easier and more accurate to make that allowance in minutes of angle on the deflection drum and aim central. If, by making an allowance of a certain number of 2-minute clicks on the sight, it is then found that the shots are slightly out of central, only a very small adjustment of aim, in addition, is needed to bring them correct.

   (iii) For aiming-off, either with the Mark 1 sight or with the aperture tangent sight, the sniper must be taught that the average width of a man's head is 6 inches, and of his shoulders 16 inches. These are the only guides he can expect on service in judging the allowance he must take. As in aiming-off with the aperture sight, great care is necessary with the telescopic sight to maintain correct elevation.

(c) *The Mark 2 and Mark 3 sights.*—On these sights small alterations, *i.e.*, clicks, of 1 minute are possible on the deflection drum. It is therefore possible to make all allowances for wind on these sights in minutes of angle, a central aim always being taken. With the Mark 2 or Mark 3 sights the sniper need never aim-off.

(d) *The Rule of Altering Sights for Wind.*—If the wind is blowing from left to right, the error is to the right, therefore, turn the deflection drum to the right (clockwise or backward). If the wind is blowing from right to left, the error is to the left, therefore, turn the deflection to the left (anti-clockwise, or forwards).

131. **Allowances for movement.**—Up to a range of 300 yards it is quite possible to hit a moving man with good regularity. Practice and the aid of the telescopic sight are needed to do this. When the target is moving across the front it is necessary to aim-off and maintain a constant lead in the direction in which the target is moving. If the rifle is checked as the trigger is pressed the benefits of the lead will be lost, and a miss behind the target obtained. For practical purposes the width of a man crossing the front may be taken as 1 foot. For a man walking at 3 m.p.h. the aim should be as follows :—

        100 yards—the man's leading edge.
        200 yards—½-foot ahead of his leading edge.
        300 yards—1½-feet ahead of his leading edge.

For a running man the above allowances must be doubled, *i.e.*, ½-foot, 1 foot and 3 feet ahead of his leading edge, respectively.

It should be noted that the " swinging through " method if engaging moving targets as taught in Small Arms Training Vol. I, Pamphlet No. 3, 1946, is not suitable for use with the telescopic sight beyond 100 yards. This is because the pointer of the telescopic sight, having no apparent width, will give less lead in passing the target than the blade of an iron foresight. Thus at ranges beyond 100 yards the shots will miss behind the target when using this method with the telescopic sight.

(B47/96) 30000 5/47 W.O.P. 27371